Climate, Agriculture and Man: Shrinking Biodiversity and Sustainability

OTHER RELATED BOOKS

Title	Author	ISBN
• Climate-Smart Agriculture Source Book	FAO	9788193698518
• Agricultural Drainage Water Management in Arid and Semi-Arid Areas	FAO	9788172334567
• Agroforestry for Climate Resilience and Rural Livelihood	Inder Dev	9789387307063
• Agrotechnology for Dryland Farmaing 2nd Revised Ed	A.M. Dhopte	9789386102737
• Changing Faunal Ecology in the Thar Desert	B.K. Tyagi	9788172334028
• Climate Change & Himalayan Ecosystem: Indicator, Bio & Water Resources	J. Sundaresan	9788172338473
• Climate Change & Himalayan Informatics	J. Sundaresan	9788172338466
• Climate Change Adaptation Strategies in Agriculture and Allied Sectors	Gslhv Prasad Rao	9788172336790
• Climate Change and Environment	J. Sundaresan	9788172338336
• Climate Change and Himalaya: Natural Hazards and Mountain Resources		
• Climate Change and its Ecological Implications for the Western Himalaya	V.L. Chopra	9788172338091
• Climate Change Impact on Ecosystem	J. Sundaresan	9788172338329
• Climate-Smart Agriculture Source Book	FAO	9788193698518
• Combating Desertification Land Degradation and Climate Change: Management of Dry Lands	T.S. Chouhan	9789388043083
• Conservation of Threatened Desert Plants	Suresh Kumar	9788172338787
• Crop Growth Simulation Modelling and Climate Change	M. Mohanty	9788172339418
• Desertification: An Indian Scenario	A.V. Rao	9788172332303
• Disease Management in Arid Land Crops	S. Lodha	9788172335175
• Drought Mitigation and Management	Suresh Kumar	9789386102393
• Drought Proofing and Water Resource Management	T.S. Chouhan	9789388172882
• Dryland Technology 2nd Ed	M.L. Jat	9788172339982
• Earth and Ocean: Useful Resources (Hindi)	D.D. Ojha	9788172339463
• Genesis and Management of Sodic (Alkali) Soils	S.K. Gupta	9789386102775
• Geo-Informatics for Combating Land Degradation	T.S. Chouhan	9789386102843
• Global Groundwater Resources and Management	B.S. Paliwal	9788172336196
• Man in the Desert	L.P. Bharara	9788172331931
• Ocean and Coastal Ecology	K. Kathiresan	9788172338916
• Organic Agriculture and Climate Change Mitigation	URS Niggli	9789383692064
• Perspective on the Nature of Geography	R. Hartshorne	9788185046525
• Physiology of Plants Under Abiotic Stress and Climate Change	A. Hemantaranjan	9788172337896
• Plant Physiology Under Changing Environment	A. Hemantaranjan	9788172336936
• Priorities of Research for Development of Arid Regions	S.K. Verma	9788172332594
• Remote Sensing and GIS Applications	D.C. Joshi	9789388172547
• Remote Sensing and GIS GPS Based Resource Management	T.S. Chouhan	9789386102867
• Remote Sensing for Natural Resources Management & Monitoring	Mahesh Gaur	9789386102720
• Remote Sensing: Principles and Applications 2nd Ed	A.N. Patel	9788172335908
• Space Technology and GIS for Disaster Monitoring and Mitigation	T.S. Chouhan	9789386102850
• Sustainable Development of Dryland Agriculture in India	R.P. Singh	9788172334055
• Thar Desert in Retrospect and Prospect	R.P. Dhir	9789387307025
• The Drought Resilient Farm	Dale Strickler	9789383692965

Climate, Agriculture and Man: Shrinking Biodiversity and Sustainability

By

Vinod K. Sehgal

Formerly Head,
Division of Entomology,
Indian Agricultural Research Institute, Pusa,
New Delhi 110 012

SCIENTIFIC PUBLISHERS

Published by
SCIENTIFIC PUBLISHERS (INDIA)
5 A, New Pali Road, P.O. Box 91
Jodhpur 342 001 (INDIA)
E-mail: info@scientificpub.com
Website: http://www.scientificpub.com

ISBN: 978-93-87893-88-7 (Hardbound)

ISBN: 978-93-87991-54-5 (Ebook)

Printed in India

Cover Photograph of Lake City of Nainital

Lake City of Nainital, popular hill station in the Indian state of Uttarakhand situated 29° N and 78°8 E, 6,000 feet altitude in mid hills of Kumaon region of central Himalayas. The city is blessed with scenic natural beauty of the majestic Himalayas, rich biodiversity which attracts large number of tourists from India and the world. Its fragile ecology is seen in patches of shrinking conifer forests, fragile mountain slopes, frequent landslides and unplanned human habitation. This place also has a living fossil tree, Ginko biloba, at the entrance to Governer's House, Nainital, Kumaon Himalayas.

Preface

Shrinking Biodiversity, the theme of this presentation is based on the perceptions that after 1940's Second world war discovery of DDT and Malthusian increase in human populations triggered by Green Revolution of early 1960's, exploitation of natural biodiversity resources increased far beyond the nature's ability to renew them. Even the quality of human resource decreased as evidenced by many fold increase in incidence of several diseases. Nature took millions of years to evolve the present dispensation of diversified flora and fauna. I have tried to show how shrinking or vanishing biodiversity at a rapid rate has affected every component of living natural biodiversity resource, including man himself. Interactions between environment and evolving populations, like competition for food, space, predation etc. are the major causes driving many species of flora and fauna to extinction.

For this study of Shrinking Biodiversity, I selected major resources of Insects, Plants and Man as together they constitute almost two third of known biodiversity resources. Few insects groups like locusts and grasshoppers which have retained primitive polyphagy, are also known to avoid several plant species. This is because insects have highly evolved senses to identify cue chemicals emanating from host-plants. Each of the diverse range of topics discussed in this write-up can be a subject of an independent treatise. To remain focused on the theme Shrinking Biodiversity, several of such topics have been kept short and precise. For too long, insects have been considered to co-evolve with their host-plants as a result of genetic feedback mechanism resulting in large diversity of secondary plant chemicals. Specific studies on internal plant feeders like leaf mining agromyzid flies have demonstrated that insect feeding alone rarely build up selection pressures to force co-evolutionary changes in their host-plants. Rather, plants evolve independently due to varied selection pressures like herbivory including insect feeding and also genotype environment interactions; insects just follow by adjusting their feeding responses by sequential evolution on closely related plant species.

Willful neglect and intervention by man into natural and agro-ecosystems together with deforestation, forest fires, urbanization etc. has caused shrinkage of natural biodiversity resources leading to unsustainability of climate, agriculture and man himself. I take full responsibility for the views expressed herein.

V.K. Sehgal

Acknowledgments

I express my gratitude to faculties, colleagues, staff, students and friends at various institutions during the last five decades of my studies and services in the agriculture sector.

I am grateful to Late Professor Dr. Brian Hocking, Chairman and Professor Dr. George E. Ball at the Department of Entomology, University of Alberta, Edmonton, Canada from whom I learnt the true dimensions of Taxonomy, Systematics and the basics of insect-host plant interactions during Ph.D. studies. I have been fortunate to learn Ecology from Dr. W.G. Evans at Alberta. With my fellow students and friends, Y.S. Krishnan, Robert Lee, Peter Kevan, A. Nimmo, J. Shorthouse, Vrinder Anand, M.S. Phul and Sharon Scheffelmaier, I shared many happy moments.

I am grateful to Dr. G.P. Sharma, ex-Head, Department of Zoology and Dr. K.N. Trehan, Professor at the Panjab University, who initiated me on leaf mining insects during M.Sc. studies. I am grateful to Late Dr. K.A. Spencer, World Authority on leaf mining flies (Diptera: Agromyzidae) and to Late Dr. G.C.D. Griffiths, Entomologist UK, during Ph.D. studies at the Univ. of Alberta for almost 5 decades of interactions on agromyzid flies.

I am grateful to Dr. Panjab Singh, ex-Director General, Indian Council of Agricultural Research, New Delhi, for an opportunity to serve at the Division of Entomology, Indian Agricultural Research Institute (IARI). I appreciate the help and cooperation I received from friends and colleagues - Drs. Y.S. Krishnan, M.S. Khalsa, Late A.K. Bhattacharya, G.C. Sachan, Deepak Ghosh, Ravinder Khanna at Pantnagar; Drs. Ramamurthy, Late A.V.N. Paul and R.K Anand at IARI and Dr. J.B.S. Bhatti, Department of Zoology, University of Delhi.

I am grateful to Dr. A. Sen, Sr. Principal Scientist at CSIR-NCL (National Chemical Laboratory, Pune, India) for many useful comments, suggestions and critical checking of the updated versions. I express my gratitude to Late Dr. Y.L. Nene, former DDG at ICRISAT and Chairman Agri-History

Foundation and Drs. M.C. Saxena. Former, ADG at ICARDA and G.K. Garg, former Professor and Head, Biotechnology at G.B. Pant Univ. of Agri. and Tech., Pantnagar for useful comments which helped me write this document. I thank Dr. V.K. Chopra, renowned Cardiologist, Director Preventive Health, Medicity, Medanta, Gurgaon, for medical advice during critical periods. Finally, I thank my dear daughter, Priya and wife Rita for their patience and support while I did this write up.

V.K. SEHGAL

Planet Earth's Largest Living Ecosystem, the Biosphere

Fig. 1. *One in Millions. Planet Earth's Largest Living Ecosystem, the Biosphere as seen from the Moon.*

The Apollo 17 crew caught this breathtaking view of our home planet as they were travelling to the Moon on Dec. 7, 1972. It's the first time astronauts were able to photograph the South polar ice cap. Nearly the entire coastline of Africa is clearly visible, along with the Arabian Peninsula.

Image Credit: NASA, *Last Updated: Dec. 7, 2015*

"You develop an instant global consciousness, a people orientation, an intense dissatisfaction with the state of the world, and a compulsion to do something about it. From out there on the moon, international politics looks so petty. You want to grab a politician by the scruff of the neck and drag him a quarter of a million miles out and say, 'Look at that, you son of a *****.'"

Edgar Mitchell, Apollo 14 astronaut, *People* magazine, 8 April 1974

List of Figures

I am grateful to Ms. **Shunhai Qu** (apjtbb@yahoo.com) dated 12/13/16 from Asian Pac J Trop Biomed for kind permission to reuse the Figures 2-6 and Fig 17 with due citation. I am also grateful to Ms. Carole Cook from U.S. EPA for use of Figs.7 and 8 vide their e mail dated 11/01/16 which states that all of the graphics and information on EPA's website are in the public domain, meaning, you may use them as you wish. A credit is appreciated (Source: U.S. EPA). All other Figures are either in public domain or are by self.

Figures 7-8. Permission from USEPA obtained

Fig.16 : Copyright/License Request permission to reuse. Permission obtained

Contents

Chapter 1

Introduction

1.1 INTRODUCTION

During the past two decades, there has been a growing realization that we as humans have depleted the original natural biodiversity resources far beyond the biological means of their retrieval. Rapidly shrinking biodiversity resources have vital linkages to sustainability of climate, agriculture and man. Large sections of almost 60% of world's population living in thickly populated countries like India, China, equatorial regions and several countries in Asia, Africa and South America are deprived of adequate basic human needs of food, nutrition, health and social security. Over 80% of world's poor live in India and these countries. This is despite massive increases in world food production due to adoption of green revolution technologies. Their life is further made miserable due to rapidly changing climate, drought, unpredictable rain fall patterns, cloud bursts, flash floods and rivers changing their course; besides sea storms, cyclones, tsunamis along the vast stretches of coastal areas. Year after year, such uncertainties lead to widespread destruction, more so in rural areas, of their homes, crops and above all loss of valuable human and livestock resources. While such events of climate shifts have been happening ever since millions of years due to dynamic natural forces that sustain Earth's equilibrium, yet the resulting damage over the past five decades or so have been far severe than ever before. The real cause is willful and systematic destruction of natural biodiversity resources. Clearing natural forests that evolved over millions of years, in equatorial regions, in tropical and temperate environments and even the vast stretches of mangroves along coastal areas are not spared to claim farmland or land for commercial and urban needs. This shrinking of forests and biodiversity resources over the past few decades increased severity of damage due to natural calamities both in hills and plains. Unplanned urban habitats without green cover, city pollution, pesticide residues in food, poor quality of health support have taken a heavy toll of human resource.

Taxonomy, systematics, inventorization of dominant natural biodiversity resources, the angiosperms and insects, together with bioprospecting of traditional knowledge of their food, nutrition and medicinal value amongst the local populations should have been one of the primary agenda of agricultural, biological and medical sciences. It takes a lifetime dedicated work to develop expertise for even a small group of biodiversity resource. We have already lost valuable expertise in several of these areas.

Dr. M.S. Swaminathan in his address at the University of Illinois in 1988 stressed the need to overcome environmental challenges, establish the symphonic agricultural systems and free the world from hunger. This is not an easy task. Amongst the environmental challenges, he cited "*Developing countries are ravaged by deforestation, desertification, demographic pressures, soil erosion and water pollution. Environments in developed countries suffer from acid rains, atmospheric pollution, contaminated water, environmental mutagens, carcinogens and toxic waste*". Amongst other areas of concern, he included loss of biological diversity, destruction of ozone layer, climate changes due to accumulation of carbon dioxide in the atmosphere; impact of ecological fires that affect our basic life-support systems, the atmosphere, fauna, flora, land and water. Almost 28 years after the above observations, many of the above components remain a formidable challenge. This book is a modest attempt to create awareness that most of the problems referred to by Dr. Swaminathan still require to be addressed urgently. This book highlights that what nature takes millions of years to create, man can destroy in less than his lifetime.

Nature evolved Man as a supreme natural resource having endowed him with ever evolving intelligence to utilize, protect, conserve and propagate this glorious heritage of living biodiversity. Man being the most recent natural biodiversity resource to set footprints on living planet earth, had this unique privilege to pass on this heritage to future generations. When first evolved, no body even in the animal world expected Man to be a saint. It was expected that he would use, plunder and even waste natural biodiversity resources. So long this use, like that of other animal species was within the biological means of rejuvenation through bio-cycling of biogeochemical resources, evolutionary diversification continued to replenish such resources.

It took several million years of evolutionary biological diversification, guided by natural selection and Darwinian struggle for survival of the fittest to arrive at the present dispensation of natural biodiversity resources. Estimates of living biodiversity indicate that we have discovered less than 2% of this precious resource. For unknown reasons, attributable only to acquired intelligence of man, we have stopped further exploration of this resource.

The manner in which sustained efforts of Alberta Bio-Innovates, Bio-Solutions and University of Alberta researches led by Dr. Stan Blade in western Canada have turned Canadian industrial hemp, banned from cultivation until 1984, into a million dollar renewable resource yielding food, fiber, plastics, medicine, nutrition and health supplement is impressive (Blade, 1998). Like hemp, nature has created millions of renewable natural biodiversity resources amongst the angiosperm flora that need to be explored for the benefit of mankind.

Nature has evolved natural remedies for all maladies facing its biodiversity resource. Many of such species have disappeared without being discovered; thus, we are losing a resource that could help provide much needed food, nutrition and health security. The United Nations led by developed countries like USA, Canada should provide renewed impetus to explore natural biodiversity resources.

At no point in time during its long evolutionary history, biodiversity resources were in short supply to meet its obligation to life and life support systems. Even when mammoth bodied dinosaurs roamed this earth during Jurassic and Triassic in the Mesozoic, there were unlimited forest resources consisting of large Tracheophytes like Filicales, Lycopsids, Sphenopsids and even Ginkgoales etc. Food was never in short supply. Amongst many causes of extinction of dinosaurs were the arrival of mammals - a lineage separated from Therapsid reptiles and birds having evolved from flying reptiles, Pterosaurs, all with better intelligence due to better brain to body ratios. During Cretaceous, most groups of vascular plants like Gymnosperms and Angiosperms had diversified which competitively displaced earlier Bryophytes, Filicales, Tracheophytes and other groups. It is believed that most groups of plants and animals are still evolving. As Man is not evolving any further, except perhaps in the level of intelligence, he has also stopped taking interest in the exploration of further diversification of plant and animal world. Another reason for this neglect could be quick successes with increased focus on cell biology and biotechnology, which prompted lesser need for study of ecology and biodiversity. The present lifestyle of Man living beyond the biological means, together with human greed for larger share of natural resources has placed serious limits on sustainability of Life, including that of Man himself. It is probable that during the next impending phase of massive extinctions, Man himself may be on the way out.

As more than 80% of small farms, with less than 1 ha holdings in developing countries as India, feed a population of 1.2 b, engage almost 60% of the population in agriculture, no price is too great to give them a better share of national productivity resources for infrastructure, sanitation and dignified livelihood wage irrespective of gender bias than what Planning Commissions have given so far. It is a matter of great concern that despite having world's best and most fertile agricultural land resources, more than 80% of the world's poor live in India. There are enough renewable and not so easily renewable natural resources to sustain present day man and natural biodiversity resources. Prevention of more than 40% avoidable food loss, between harvests to mouth is also a matter of global concern. Freedom from hunger and poverty has vital linkages where the debt trapped small farmer is lured to by a combined nexus between pesticide companies, pesticide dealers and commission agents who exploit farmer miseries by extending credit with hidden costs for a long list of seed, chemical fertilizers, weedicides, pesticides and even growth hormone applications. In fact, his crop is sold out even before sowing and harvest leaving him with a begging bowl at the mercy of lenders. In the last 5 decades, the country has witnessed an increasing spate of farmer suicides; however, no pesticide company, pesticide dealers and in between commission agents has seen any dip in their profits. In a free for all unmonitored environment, people are consuming ever increasing chemical and pesticide residues, breathing polluted air, facing wide spread apathy to infrastructure neglect, poor hygiene, sanitation woes and unsustainable living habitats. Several studies have linked pesticide residues and pollution to massive degradation of human health, evidenced by increasing incidence of diseases affecting immune systems, cardiac and respiratory problems etc.

Over the past 5 decades, sustained efforts of Indian Council of Agricultural Research (ICAR) and its affiliated institutions, the State Agricultural Universities (SAU), International Agricultural Research Centers, IAASTD and SAG efforts of UNEP and World Food Program (WFP) of FAO Rome has generated enough knowledge and technologies to produce healthy, nutritious food. Researches have shown that moderate productivity gains are possible by delinking HYV's from costly inputs of toxic pesticides, chemicals in standing crop and by retaining effective seed treatment, sustainable use of renewable natural resources, initial booster dose of organic/farmyard manures, water, minerals etc. for healthy plant stand. Biodiversity conservation and sustainable development are two inter-related branches focusing on social progress, economic growth and environmental protection on one side, and ecosystem conservation on the other. International initiatives of UNEP and the newly formed 17 SDG's (Sustainable Developmental Goals) fully integrate social, economic and environmental dimensions of sustainable development goals. It has suggested ways and means to tackle poverty and help farmers and World agriculture, produce wholesome healthy food with due regard to environment, to fight malnutrition in humans. It indeed is a great hope mankind is looking for. This world, particularly developing countries will need to allocate resources to meet the challenges set by the newly adopted Sustainable Development Goals.

1.1.1 Nature: A Tribute

"Sitaron se aagey jahan aur bhi hai,
Abhi Ishaq ke imtihan aur bhi hai."

(There are worlds beyond the stars we can see,
For this passion we must pass many more tests)

There is so much yet to be discovered that makes us treat Mother Nature as a supreme natural force, which evolves life, creates variability and selects the ones fit to survive. It has created terrestrial, aquatic, freshwater and marine habitats and ecosystems, turning planet earth into a self-perpetuating biosphere, unique amongst the galaxies, Milky Way and beyond. Most civilizations acknowledge this supreme power that has given the most beautiful and bountiful world of natural, physical and chemical resources that support this glorious heritage of living biodiversity.

The recent discovery of Boson, Higgs particles or God's particles as the finest particles of matter, that can combine under special circumstances to create new matter, simple molecules and compounds and even biomolecules which may become precursors of life. As laws that govern the formation of matter, natural, chemical and physical resources do not vary across planets, galaxies, milkyway and beyond, it seems more likely than ever before, that another planet with similar history as on earth which supported spontaneous creation of life under unique set of circumstances could be discovered soon. Presence of water, favorable temperature regimes, atmosphere and environment would be the prerequisite. If it happens, it seems logical to assume evolution of life, guided by natural selection would also have proceeded on similar lines as on earth. There could be microbes, diverse life forms, mammoths and even more intelligent forms of life than on earth.

Chapter 2

Life on Planet Earth

2.1 EARLY CLIMATE, ORIGIN OF LIFE AND EVOLUTION OF CONTINENTS

It all began over three billion years ago. Though formation of biomolecules happened during this period, yet cellular life, as *Amoeba* was nowhere. How the first forms of life were created and how it evolved? There are plenty of hypotheses and logical conclusions based on available evidence, yet no fossil remains are available to support them.

2.1.1 Planet Earth: The Beginning

Planet Earth, created out of Big Bang about 10 billion years ago, took almost 7 billion years to cool and evolve the environment that would favor creation of life. There was no free oxygen or nitrogen in atmosphere. There were plenty of water vapors, ammonia, methane and some carbon dioxide. This was followed by long Precambrian period, beginning 3,000 million years till about 600 million years, the beginning of Paleozoic Cambrian. Much of the earth surface was covered with vast stretches of hot water oceans. There was probably intense volcanic activity that also led to the formation of mountain ranges and deep ocean valleys.

Terrestrial land took shape, with firming up of tectonic plates and formation of lithosphere. Most of the land was contiguous. As there was no protective ozone layer, passage of strong ionizing UV radiation from the sun could have produced organic compounds, including amino acids in shallow coastal seas and other water pools. Concentration of such compounds in ocean could have produced large bio-molecules, even nucleotides like DNA, which when energized in enabling medium could self duplicate. Formation of such molecules as viruses and bacteria could not be regarded as living because they required organic living medium to perpetuate.

On planet earth, it must have been a long journey to evolve simple forms of heterotrophic unicellular life as an *Amoeba*. It took further several million years to evolve the rich diversity of autotrophic green plants and heterotrophic animals through interdependent co-evolution or sequential evolution between green plants and animal species, hosts and parasites, biological species and habitat etc., besides mutations, polyploidy and inheritance of environmentally acquired neo-Lamarkian attributes.

2.1.2 Origin of Life

Most biologists agree that life originated spontaneously from non-living matter, under unique set of environmental conditions that once prevailed sometimes during the long stretch of Precambrian, beginning about 3,000 million years ago. Evolution of unicellular heterotrophs like *Amoeba* with nucleated protoplasm having innate energy regulation to reproduce another living cell of its kind must have been a long and gradual process. It is possible that the precursors of amoeboid form of life were non-nucleated unicellular heterotrophs, which multiplied like amoeba by division of cytoplasm.

A full discussion about the formation of cell membrane by bonding of lipid and protein molecules, cell theory, origin of cytoplasm with different organelles like nucleus, mitochondria, endoplasmic reticulum, Golgi bodies etc., are subjects too large to be included in the theme of shrinking biodiversity. Some scientists believe that the mitochondria, which have independent DNA from that of nuclear DNA were probably once free living organisms with their own energy regulation, before they were gobbled up by the cellular cytoplasm of amoeboid heterotrophs. Ever since then, they probably survived mutually beneficial relationship.

Probably cytoplasmic inheritance of cell inclusion bodies like pigments, mitochondria, chromatids and even chlorophyll played an important role in evolution of early organisms. Cytoplasmic inheritance, though not fully appreciated, still plays an important role in genetic inheritance. Absorption of ammonia and methane and other organic molecules by such organisms increased the level of carbon dioxide and nitrogen in atmosphere.

Rapid evolutionary diversification of such heterotrophs, formation of mitochondria like cell inclusions for oxidative release of energy, transformed ecological conditions, paved the way for the origin of the first autotroph with light sensitive pigments to absorb visible light as the source of energy. Such pigments perpetuated largely through extra chromosomal cytoplasmic inheritance that evolved into more efficient chlorophyll. Early photosynthetic organisms and plants, mostly protists and algae, utilized carbon dioxide and increased the concentration of oxygen in atmosphere. This oxygen took several million years to accumulate and form the protective ozone layer that filtered most of the UV, but did not absorb visible light. Origin of chloroplasts in unicellular organism like *Chlamydomonas* and algae marked an early division into autotrophs and heterotrophs. This led to diversification of autotrophic phytoplankton flora and algae and heterotrophic zooplankton, and multicellular fauna.

After evolution of simple autotrophs and heterotrophs, transformed ecological conditions by then were very different from those that made spontaneous origin of heterotrophs and simple autotrophs possible. Thereafter, life diversified entirely through evolutionary processes, which supported multicellular life forms. This also laid the foundation of reproductive perpetuation of life. Although life has not yet been created in the laboratory, it is probable that with recent progress in biological sciences it may become a reality sooner than we think.

2.1.3 Planet Earth: From Single Continent Pangaea to Continental Split

A single supercontinent, called Pangaea existed during the late Paleozoic and extended until the late Mesozoic eras. The name Pangaea is derived from ancient Greek *pan* meaning "entire", and *Gaia* meaning "earth". The single global ocean, which surrounded Pangaea, is accordingly named Panthalassa (Wikipedia, 2012 a, b, c). Pangaea broke into two pieces, one in the north, the continent of Laurasia and the other in the south, the Gondwanaland (Wikipedia, 2012 e, f, g). Till about two hundred million years ago, the continents of Antarctica, South America, Africa, India, Australia and Zealandia were joined together as a single supercontinent known as Gondwana, or Gondwanaland (Fig. 2). Joseph (2006) also reveals the existence of land called Lemuria in north, which separated from Laurasia 200-180 million years ago during the breakup of Pangaea (Fig. 3). It formed one of the world's oldest civilizations. Later, due to tsunami, people regularly moved from the lost continents of Island Mu to Island Atlantis in Mexican Sea and to Kumari Kandam, the lost continent that existed south of present day Kanya Kumari district in Tamil Naidu at the southern tip of India, in Indian Ocean, about 100,000 years ago

(Wikipedia, 2012 d) (Fig. 4, 5, 6). There are references in Tamil Sangam (Sangam literature) classics to a landmass where people lived and which was later swallowed up by the sea. The literary references of the lost land were about three Tamil Sangams, the Academy of poets also known as assemblies of Siddhars who were the great scientists of ancient days.

The first two Sangams took place in a sunken continent, Kumari Kandam with the first Sangam located in Thenmadurai that lasted for 4,440 years (Wikipedia, 2012 e, f, g) (Fig. 6) and had 4,449 poets with 549 members including the head of the Sangam, Agathiyar, also referred to as the father of Siddha medicine. Agathiyar specialized in medicine, language, alchemy and spirituality. After the sea swallowed Thenmadurai, the capital was shifted to Kaptapuram. The second Sangam functioned for 3,700 years and had 59 members. This city was also submerged in the sea. The capital was then shifted to Madurai in the peninsular India. Here, the third Sangam was established and functioned for 1,850 years. This academy with 49 members and 449 poets were described as having participated in the Sangam. The Tamil Sangam classics refer to a populated land Kumari Kandam, 100,000 years BC to 50,000 years, which was submerged in the sea.

2.1.4 Continental Drift

Continental drift being a relatively recent event in late cretaceous, about 200 m years ago, almost all major groups of plants and animals were fully evolved by then. Differences in present day distribution of flora and fauna indicate diversifications after the continental drift started.

Fig. 2. *A single supercontinent called Pangaea existed from early Paleozoic uptil early Mesozoic era*

Continental drift refers to the movement of Earth's continents relative to each other. The speculation about continental drift was first formally developed by Alfred Wegener in 1912 but it was largely rejected due to lack of evidence. This view has now been replaced by the theory of plate tectonics, which describes large-scale movement of Earth's lithosphere. The lithosphere is the outer most rigid shell of planet earth, which is broken up into tectonic plates. The theory of tectonic plates and related developments had been discussed by Dietz (1961) and Blacket et al. (1965).

Continental drift occurred in the late cretaceous, almost 200 million years ago, when a single super-continent Pangaea rotating anti-clockwise split into northern Laurasia and southern Gondwana regions. The northern Laurasia included the present China, Europe, North America and Asia. The southern Gondwana included India, Africa, Madagascar, Antarctica, Australia and South America. Later, the Gondwana also split as

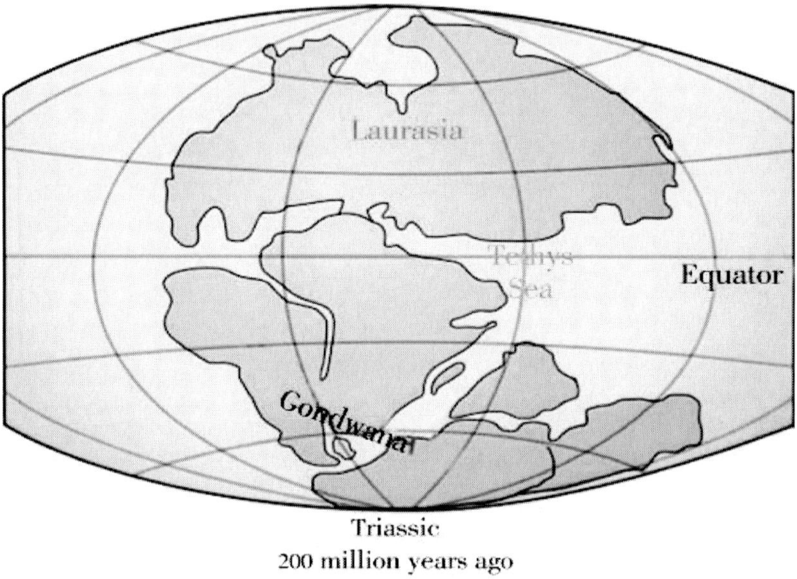

Fig. 3. *A single supercontinent Pangaea splits into northern Laurasia and southern Gondwana during Triassic*

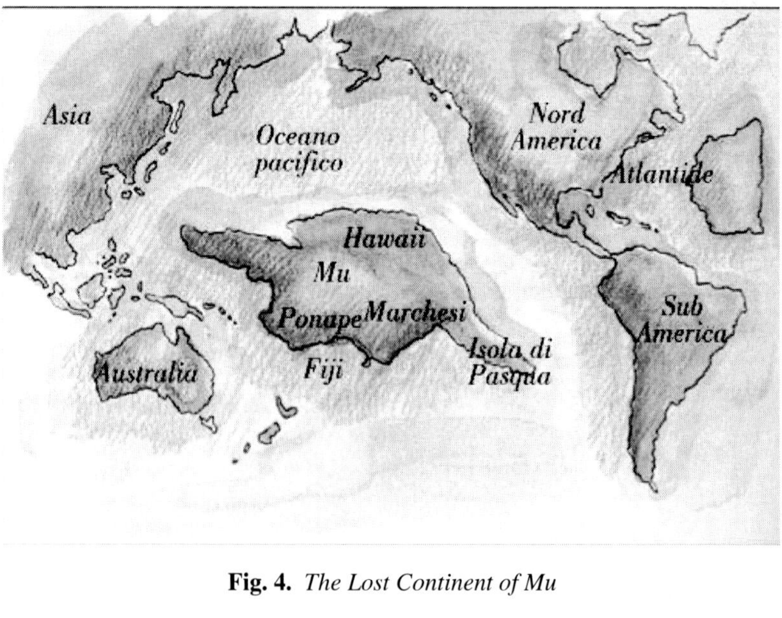

Fig. 4. *The Lost Continent of Mu*

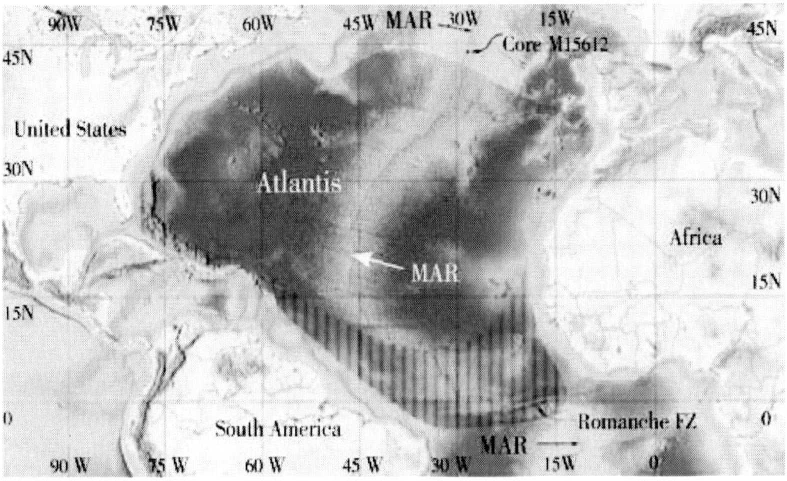

Fig. 5. *The Lost Continent Atlantis*

Fig. 6. *The Lost Continent Kumari Kadam, south of Kanya Kumari, India existed from 100,000 years BC, for about 50,000 years*

Antarctica, Madagascar, India and Australia separate from Africa. South America drifts further westward. India drifts northwards. It was a long journey for India from southern hemispere before it crashed into the Eurasian Plate in the northern hemisphere. There were no Himalayas. There were ferns, mosses and conifers all around. This is when some dinosaurs roamed. The relative movement of the tectonic plates may be convergent, divergent or transformed. Earthquakes, mountain building like the Himalayas and oceanic trench formations occurred along these plate boundaries as the continents drifted. All major groups of plants and animals including the dinosaurs were fully evolved by then. As the continents drift they take their flora, fauna including their dinosaurs with them. The relative movement of these plates ranges from 0 to 100 mm annually. Almost 65 million years ago, an unusual happening, a meteorite strike or a major volcanic eruption occurred, which wiped out almost 80% of the flora and fauna. Most of the dinosaurs were also wiped out. Birds however escaped extinction. Leftover dinosaurs survived for some time on the shrinking biodiversity resources and perished later due to competition. Almost 40-50 million years ago, the Indian plate crashed into the Eurasian plate, creating the Himalayas. The story of the rise and extinction of Dinosaurs has recently been published in Indian Newspaper, Hindustan Times, New Delhi, (HT, 26 March 2017), credited to Rachel Lopez (rachel.lopez@hindustantimes.com). The report highlights that almost 1900 species of dinosaurs have been discovered so far, out of which almost 20 species are from India.

Recently, a team of Scientists from New Zealand, New Caledonia and Australia, led by Mortimer et al. (2017) in a Journal article published in GSA (Geological Society of America) TODAY have concluded that New Zealand and New Caledonia, which, till recently were considered as part of Australia, are a part of the separate new eighth continent **Zealandia**. This was part of the southern landmass called Gondwana from which it separated around 85 million years ago. Almost 94% of this continent now lies submerged in ocean. This discovery may explain the distinctness of the fauna and flora of New Zealand.

2.2 NATURAL RESOURCES

During the long history of evolution of biological diversity through geological time and space, nature had also evolved and accumulated rich wealth of natural resources, both renewable as well as those not so easily renewable. Planet Earth has never been resource poor or resource deficient to sustain and nourish living biodiversity resources.

2.2.1 Non-Renewable and Renewable Resources

Vast reservoirs of renewable resources that can benefit mankind lie untapped in natural biodiversity of plant and animal world. Crop residues, plant waste, harvest from forests etc., if recycled can be turned into food, fiber and organic manures. Several plants amongst the temperate flora like *Achillea, Artemisia, Solidago, Urtica, Tanacetum, Lantana, Tephrosia, Cajanus,* which grow wild in Alberta Rockies and in Indian Himalayas are known to yield, fiber food and medical supplements. In addition, plants contain wealth of bioactive molecules in the form of secondary plant metabolites like terpenes, alkaloids, phenolics, insect growth hormone mimics etc. These substances form the plant's natural defense mechanism against herbivores. Many of such substances have toxic effects and are insect repellents, antifeedants etc. which can be used to control harmful insect populations.

Priyanka Das, Pune Mirror (Oct 19, 2016) reported that at a time when Pune, India is riddled with dengue cases and mosquitoes are increasingly becoming resistant to market repellants, grapefruits have come to the rescue. City scientists have synthesized a compound to tackle the dengue causing *Aedes aegypti*, a mosquito that is also a carrier of the Zika virus, from elements that are naturally found in grapefruits. The compound developed is comparable to DEET (N-diethyl-meta toluamide), a chemical compound that was developed in the USA in the 1940s and is found in many insect repellents in North America and Europe. DEET is considered the golden parameter when comparing insect repellents, shared Avalokiteswar Sen, Entomologist, Organic Chemistry Division at the CSIR-National Chemical Laboratory, which formulated the new repellant.

Recent surge in the use of herbal and Ayurveda products for human welfare is a step in the right direction in the use of renewable natural resources. As long as demand and exploitation of these resources is met through biocycling of biogeochemical cycles, these will never be in short supply.

Mineral resources can also be sustained for longer time, if demographic demand on such resources is partially compensated through recycling of used materials (Cloud, 1969; Odum, 1971). Thus, conservation and judicious use of such resources within biological means assumes significance for sustainable development of living resources.

All living organisms constantly interact with non-living chemical and physical resources and with other living natural resources like forests, rivers and marine ecosystems to establish energy flow to sustain life. Natural resources like iron, coal, oil and other minerals are non-renewable as it takes millions of years to replenish, while others like energy, living resources, forests etc. are renewable.

2.2.2 Water

As water constitutes more than 90% of every living cell, it is the most important resource to keep the biological clock ticking. Rainfall patterns and total precipitation determine vegetation patterns worldwide. Tropical and semi-tropical rain forests support a huge variety of broad leaf trees that yield rubber, coffee and sandalwood, among precious natural resources besides also providing habitat to a rich diversity of biological natural resources.

In semi-tropical areas, water can also be a severe limiting factor when it is not evenly available for human and agricultural activity. Biodiversity, especially of forests help retain soil moisture recharged during surplus rains and its slow release during lean periods. Cropping systems which do not drain large quantities of water are best suited to such regions. Water stress, if not adequately addressed through soil moisture recharge during periods of uncertain rainfall can become a serious limiting factor for crop growth. In most other habitats, like in monsoon dependent regions of northern plains in India, rainfall occurs in a narrow span of one to two months of July and August. Despite being water surplus region, most water is lost through runoff or through rivers, often leaving an intermittent dry spell. Annual monsoon rainfall, which should indeed be a cause of cheer, happiness and jubilation, can also cause misery. Surplus water runoff, if not harvested through depleted forests and proper watersheds, can bring along with it huge quantities of loose soil and gravel which has already silted most rivers and the downstream dams. Rivers often change their course; floods wash off many villages and large stretches of agriculture land. Many lives are lost due to this every year. Both man and his agriculture suffer untold miseries year after year. The political response is often limited to making army/state helicopter surveys of flood affected areas, order air drop of food packets, press a few boats to rescue, a small compensation is also announced. Such window dressing approaches do not go very far. The issue is closed for another year till monsoon comes again.

Water is widely regarded as the most essential of natural resources, yet freshwater systems are directly threatened by human activities (UNESCO, 2009; Vorosmarty et al., 2005). Vorosmarty et al. (2010) observed human and biodiversity perspectives on water security using a spatial framework that quantifies multiple stressors and accounts for downstream impacts. They reported 80% of the world's population is exposed to high levels of threat to water security. They suggest that it is necessary to limit threat to global water security for both humans and freshwater biodiversity at source rather than resort to costly remedial measures.

In marine coastal habitats, depleting mangroves due to mining and commercial activities, in many countries and along the Western Ghats and Bay of Bengal in India, has endangered the rich biodiversity of marine fauna and flora. Unsustainable harvest of marine biodiversity and pollution due to toxic run off and leakages of oil during transportation are common.

Deforestation in urban habitats together with shrinking of biologically diverse original forests in the nearby mountain ranges, ecologically unsustainable practice of burning understory in forests by the forest department to make easy approach roads, altogether make it almost impossible to rejuvenate such forests.

Rain water harvesting, creating a network of water sheds in hill slopes and in plains, linking of major rivers to form large reservoirs are some possible elite water management strategies often discussed every year, but little or no implementation occurs on ground.

2.2.3 Soils

Biodiversity resources help retain and bind organic matter in alluvial soils of north Indian plains. This helps restore fertility after crop harvest. Recycling of biological waste after crop harvest also enhances soil fertility. Undue demand on ground water resources by crops like summer paddy (rice) lowers ground water table and brings salt to surface, as has happened in vast tracts of north western plains in India which makes reclamation of such soils a difficult task. Roadside plantations of *Eucalyptus* in such areas further lowers water table. This has already led to spread of arid desert conditions in the northwest plains of India. In most parts of peninsular India, both alfisols and vertisols are relatively poor in organic matter. Thus, yields of field crops like paddy (rice) and pulse crops are almost one half of those in northern plains. Suitable adjustments in cropping systems, introduction of value added horticulture and dairying might bring the desired change to restore some of the shrinking biodiversity resources.

2.2.3.1 Soil Moisture Recharge

Uncertainties associated with soil moisture recharge during the rainy season due to *El Nino* effects will affect agricultural productivity particularly in semi-arid regions of Asia, Africa and South America. Saxena (2015) at CGIAR Institute and ICRISAT has highlighted the need to harvest every single drop of water to mitigate the sufferings of small farmers and help eradicate poverty. Saxena (2015) advised *in situ* water harvesting by constructing micro- and macro- catchment area, assisted by development of innovative Laser-guided Vallerani plough technology by designing the location and size of the rainwater collecting structure in a watershed using GIS and remote sensing technologies.

Chapter 3

Climate Change and Environment

3.1 CLIMATE CHANGE AND ENVIRONMENT

The adverse impact of shrinking biodiversity on changes in climate and environment has been severe due to clearing of tropical rain forests 10 degrees north and south of equator and in subtropical regions for agriculture and urban needs. Clearing vast stretches of mangroves for commercial activity, uncontrolled forest fires in temperate North and South America, Europe and Asia have contributed to shrinkage of natural biodiversity. These together with unsustainable burning of fossil fuel, oil and coal, emission of greenhouse gases and rising temperatures over the Pacific Ocean are all links of the interconnected chain that causes unpredictable shifts in rainfall patterns across Asia and other equatorial regions. Similar views were expressed for Africa (Brown, 2015). He stated that the underlying issues that these processes aim to address climate change, sustainability and development are not independent, and success in each is likely to be highly dependent on success in the others. The intensity of these events has increased several times since after 1960's increase in demand for agricultural land due to green revolution crops and for urban needs.

A web portal developed by Late Dr.Y.L.Nene, Chairman, Asian Agri-History Foundation, India, (www.asianagrihistory.org, 1994) has an excellent coverage for subjects like climate, environment, biodiversity, agriculture etc.

3.1.1 Global Warming

Depleting biodiversity, forest fires, unprecedented emission of carbon dioxide and other industrial pollutants has led to unsustainable increase in summer temperatures, triggering fears of global warming. Such increases are significant in tropical and semi-tropical countries. This can have far reaching and disastrous consequences on sustainability of remaining biodiversity resources and survival of man in affected areas.

To be forewarned is to be forearmed. Global Green House (GHG) gas emissions due to human activities have grown with an alarming increase of 70% between 1970 and 2004 (Singh, 2007). Increase in the concentration of these gases increases Earth's surface temperatures. Green house gases that trap earth's heat are accumulating far beyond the ability of forests and earth's ecosystem's ability for carbon sequestration, a natural or artificial process by which carbon dioxide is removed from the atmosphere and held in solid or liquid form. 'Climate Change Report' of IPCC 2007, claims that the consequences of GHG *vis-a-vis* global warming would last for a century even if carbon dioxide emissions are stabilized now. To prevent further damage, time to make amends is now. Climate change affects the poorest people the most. They may need to bear with severe droughts, strong winds, devastating floods, insecure livelihood etc. Climate change affects food production due to temperature fluctuations, precipitation etc. that affects crop yields.

For planned policy and preventive measures, setting a time frame, *i.e.*, year in which ongoing global warming will affect Earth's biodiversity is necessary. Mora et al. (2013) suggested study of climate variability, instead of overall climate means to understand climate change. With this rational approach, they tried to identify the year in which climate change will cross the thresholds, whereby events that are currently regarded as extremes will become the norm. Huntingford and Mercado (2013) advocated setting time frame, potential year as 2050, to refocus policy debate and for action when significant climatic change will impact Earth's ecosystem and biodiversity. Post (2013) has suggested urgent steps to save tropical biodiversity where tolerance to climate variability is narrow. Such tropical species will face extinction sooner than we may expect.

A recent study Global Warming Forecasts - 2015 undertaken by climate experts in USA, UK and China and also supported by Indian Agricultural Research Institute has estimated 8 degrees Celsius rise in summer temperatures in most parts of northern India (HT 15 July 2015, India).

Estimated summer temperatures may cross 50° Celsius by the end of this century. This will not only restrict outdoor human activity but will impact agriculture and human health adversely. Together with diminishing resource of fresh water, it could make life even more miserable for man and his life support systems on earth. Though increases in sea temperatures will not be as marked, but such rising temperatures are bound to increase sea levels due to melting of polar ice caps, endangering lives of many countries in the equatorial region. Sea levels are sensitive to global warming. Warrren Cornwall, a freelance journalist in Washington in his post published in Oceanography, Feb 22, 2016 has reported about four papers published in *Proceedings of the National Academy of Sciences* which suggests that sea levels are rising at their fastest rate in 2,000 years due to global warming. "*The big takeaway is that the modern rate of sea level rise in the 20th century is faster than anything we've seen in the previous two millennia*" says Benjamin Horton, a geologist at Rutgers University, New Brunswick, in New Jersey who helped direct one of the studies. "*This isn't a model. This is data.*"

3.1.2 Change in Climate, Biodiversity and Greenhouse gases

Climate change refers to the variation in the Earth's global climate or in regional climates over time scales ranging from decades to millions of years. A Report on Climate Change: Challenges to Sustainable Development in India; (Occasional Paper (3). 2008), Rajya Sabha Secretariat, New Delhi prepared by Prof. M. S. Swaminathan, MP, Shri N.K. Singh, MP and Dr. R. K. Pachauri, Chairman, IPCC and Director-General, The Energy and Resources Institute (TERI) has given valuable suggestions for the benefit of science and policy makers in India. Climate change has been attributed directly or indirectly to human activity that alters the composition of the global atmosphere and which is in addition to natural climate variability observed over comparable time periods. This report concluded that climate change has the potential to cause immense biodiversity loss, affecting both individual species and their ecosystems that support economic growth and human well-being. It is difficult to predict the overall result of climate changes on animal and plant kingdom. Devastating effects on the native habitats of many animals and plants due to global warming is likely to drive a considerable number of today's known animal and plant species to extinction. Mass extinctions of the Earth's flora and fauna have occurred before also but those were driven by natural factors. However, the projected extinctions of flora and fauna in the future will be human driven *i.e.*, due to adverse impact of human activities.

Linkages between climate and biodiversity influence wide geographical areas irrespective of political borders. The manner in which *El Nino* affect warming of temperatures over the Pacific, has started affecting rainfall patterns in Asia over the last decade assumes significance due to rapid loss of Tropical rainforest biodiversity. Together with uncontrolled forest fires and increase in carbon dioxide emissions and industrial pollution are probably a major reason for global warming. Greenhouse gases, which include carbon dioxide, methane, nitrous oxide, chlorofluorocarbons, hydrofluorocarbons, sulphur hexafluoride etc. increase surface temperatures. Water vapours and ozone normally absorb and emit solar radiation within the thermal infrared range. Greenhouse gases, if not recycled by natural forests can damage the protective ozone layer. In the absence of cloud cover, greenhouse gases and atmospheric pollution affect the ozone layer, thereby resulting in larger amounts of solar radiation reaching the earth's surface. If greenhouse gases continue to increase, earth's temperatures would rise to levels where they would be harmful to biodiversity and people worldwide.

Two major contributors of carbon dioxide emission are use of fossil fuel -57%, and deforestation - 17%. Other gases are methane - 14%, nitrous oxide - 8% and flourinated gases - 1% (Fig. 7). Country wise break up shows China, USA and European Union contribute 50% of greenhouse gases, followed by India (5.7%), Russia (5.4%) and the rest by other countries (Fig. 8). Despite the fact that India's contributions to greenhouse gas emissions are very small; the Government of India has taken many measures to improve the situation in this regard. The Ministry of Environment and Forests is the nodal agency for climate change issues in India. India has initiated several climate-friendly measures, particularly relating to renewable energy. It has one of the most active renewable energy programs besides having perhaps, the only dedicated Ministry for Non-Conventional Energy Sources in the world (Ministry of New and Renewable Energy).

In its Fourth Assessment Report, the Intergovernmental Panel on Climate Change (IPCC) states with '*very high confidence*' that human activities since 1750 is the major cause of worldwide trends of global warming. For the next two decades, says the IPCC, about 0.2 C of warming per decade is projected. Among the many knock-on effects are a sea level rise of more than 50 cm by 2100. It is also likely that between 20 and 30 per cent of plant and animal species assessed will be at greater risk of extinction if the rise in global average temperatures exceeds 2-3ºC (IPCC, 2007; Fischlin et al., 2007).

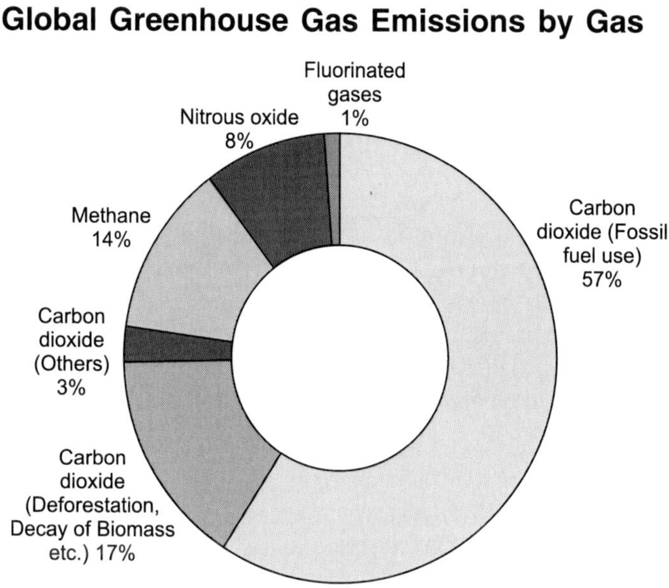

Fig. 7. *Global Greenhouse Gas emissions*

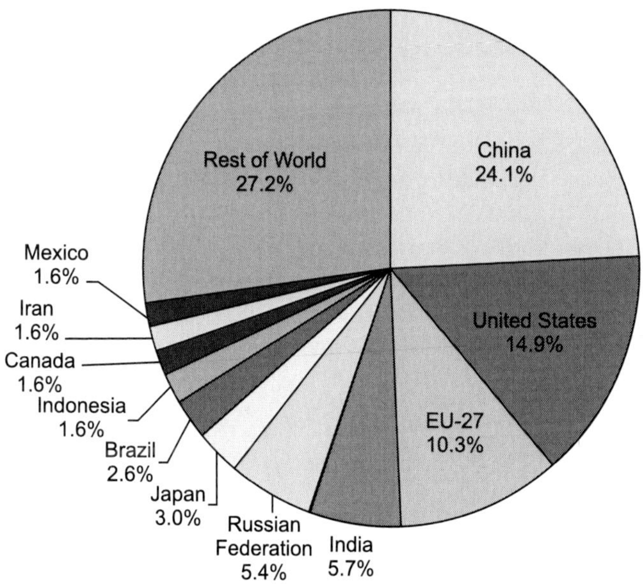

Fig. 8. *Global Greenhouse Gas Emissions country wise*

3.1.3 Volcanic Eruptions, Floods, Cloud Bursts, Sea Storms

All natural calamities like volcanic eruptions, earthquakes, floods, cloudbursts, sea storms, tsunamis in coastal regions are a natural phenomenon occurring since several thousand million years. Human concern is hardly ten thousand years, when major civilizations evolved and adopted agriculture to produce food. However, the gravity of their impact and loss of human life is clearly compounded by deliberate human abuse of natural resources like unsustainable mining of natural resources, deliberate deforestation for unsustainable urbanization etc. Similarly, clearing mangroves along coastal areas for commercial activity had adversely affected life of those who depend on marine resources for livelihood.

Adverse effects on human life and environment have been severe since after the Second World War, incidentally, more so after the use of DDT in agriculture and adoption of green revolution crops since late 1950's. Rapid rate of increase in human populations due to increased food availability since 1960's led to unsustainable exploitation of natural resources like clearing of forests, mining of renewable and not so easily renewable mineral and land resources, clearing of mangroves, unsustainable harvest of marine biological resources etc. These, together with forest fires and unsustainable human habitations along major river valleys, burning fossil fuel and coal etc. has produced greenhouse gases beyond nature's means of their biological cycling. It is this damage of natural resources that has resulted in shrinking natural biodiversity resources. So, why blame nature for land slides, uncontrolled river waters damaging human habitations, silting of rivers, dams, rainy season flooding of villages, loss of field crops and above all general degradation of human health, nutrition and social security? Lack of ecologically sustainable natural resource management policy including viable forest management policy, unplanned expansion of urban habitats without green cover, together with road side dumping of urban waste, poor infrastructure, untreated discharge of industrial pollutants and carbon dioxide emissions, toxic suspended particle matter etc. have made life miserable and living ecosystems unsustainable.

There are positive aspects that are also associated with nature's natural calamities. These include renewal of depleted top agricultural soil with minerals necessary to sustain crops and biodiversity resources. Even volcanic eruptions bring to the surface huge amounts of iron, calcium, magnesium, and phosphorus which are necessary to sustain biodiversity resources.

Chapter 4

Understanding Biodiversity

4.1 BIOLOGICAL SCIENCE: BASICS AFFECTING BIODIVERSITY

As sustainability of human life, biodiversity, agriculture and forestry depend almost entirely on applications of biological science to life systems, it is necessary to review them briefly.

4.1.1 Biocommunications: Birds of Feather Flock Together

It is a common sight to see communities of invertebrates, insects, birds, mammals and plants to keep their flock together. Biocommunication between members of species populations has been the key to survival and propagation of their flock. Evolved over millions of years of bio-diversification in response to Darwinian natural selection, these have diversified into several kinds of adaptations for harmonious coexistence and propagation of communities. Members of the species populations in an organized community actively work for mutual welfare and defend the space they occupy. Several of the social insects, communities of birds, other animals and even plants are an outstanding example of this. We, as humans still have a lot to learn from them.

Several species populations have common requirement for environment or living space. This leads to interspecific competition. Lotka (1925) and Volterra (1926) had developed a mathematical model, independent of each other for survival of competing populations which indicated that only one species population will survive in a particular niche environment. This principle is often referred to as Gause's (1934) principle of competitive exclusion based on his studies on competing populations of *Paramecium* species. Since then, similar results have been obtained for competing populations of several other animal and plant species. Gause's principle of competitive exclusion has ensured specialised space or niche for every species population. This has led to the development of a specialized field of study relating to community ecology as a means of population regulation. Several types of such interactions are briefly referred to in the following paragraphs.

4.1.1.1 Intra-specific Interactions

Pheromone trails in ants is a means of exploring environment for food and security. The waggle dance or wag tail dance performed by honey bee workers returning from a food source located more than 150 m to guide foraging workers is an extra ordinary example of intraspecific biocommunication. Classical studies by Von Frisch discovered the mystery of such honey bee dances which earned him the Nobel Prize for Medicine in 1973. Several studies have also demonstrated that floral odours on the forager's body also play an important role in this communication. Auditory signals in mating calls of crickets, cicadas, amphibians and birds; visual signals of lantern beetles also come in this category. Cannibalism does occur in some animal species populations as seen in preying mantis, where the female after copulation may devour the male.

4.1.1.2 Inter-specific Interactions

Symbiosis, mutualism and commensalism as mutually beneficial associations are common in animal world. Parasites and predators have emerged as important means of population regulation. Parasites are smaller than the host and rarely kill the host, whereas predators destroy the host. Insect-plant relationships and parasite-host relationships, discussed later have important implications in agricultural, animal and medical sciences. Antagonistic fungi, like *Trichoderma* and *Gliocladium* play an important role in the control of harmful pathogenic fungi. Such components of biodiversity when occurring in their natural habitats play an important role in sustainable developments. Many of such interactions include the nitrogen fixing bacteria *Azobactor* and legume hosts that allow fixation of atmospheric nitrogen in root nodules for healthy plant growth. Secondary plant chemicals in *Urtica, Parthenium* etc. inhibit the growth of other native vegetation in their vicinity. Territoriality exhibited by several of the animal species has emerged as an important means of survival in hostile environments. Alien species, as referred to the introduced species like *Lantana camara, Parthenium hysterophorus* in India produce allelopathic chemicals that result in competitive displacement of native vegetation over large areas of their new distribution.

Attractants and repellents released due to interactions between several species populations, particularly insects and environment, vertebrates and environment, plants and environment that evolved over millions of years have guided coevolution and/or sequential evolution as a genetic feedback mechanism. These interactions between plants, pollinators and other herbivores facilitate seed setting and dispersal to new areas. Such signals also trigger aggregation and migratory responses in species populations. Late Dr. M.S. Mani, an eminent authority on parasitic hymenoptera, Chalcidoidea and plant galls working at St John's College, Agra, India published several articles and books relevant to biodiversity and adaptation to special environments (Mani, 1962, 1964, 1968, 1974 and 1978).

Evolutionary and phylogenetic implications of insect, plant and environment interactions based on biogeography, sensory physiology of various receptors are beyond the scope of this write up. Vast amount of literature, classic and recent is available on such subjects (see Fraenkel and Gunn, 1940; Dethier, 1947; Whittaker and Feeny, 1971; Schoonhoven, 1972; Ananthakrishnan and Sen, 1998 etc).

4.1.2 Evolution: Descent with Modification

The general acceptance of the synthetic theory of evolution during the twentieth century as production of variations, and natural selection of ones fittest to survive and reproduce has enabled a holistic view of the entire process of biological diversification.

Publication of Charles Darwin's (1859) '*Origin of Species*' highlighted the importance of geographical variations and diversification of biological species populations. Later, wide acceptance of Lamarck's (1815) views on hereditary effects of environments and inheritance of acquired characters and de Vries (1909) mutations and polyploidy became part of the long and gradual process of origin and diversification of biological species. Brilliant discussion of these can be found in Simpson and Beck (1965), Mayr (1963) and Stebbins (1950).

4.1.3 Genetics and Inheritance of Characters

Discovery of laws governing inheritance of characters in plant and animal species (Mendel, 1866) now explained the natural variability in species populations. Laws of dominance and segregation of alleles laid the foundation of modern genetics. It also led credence to gene concept and evolution of species as evolving gene pools that created variability and biodiversity of species populations. The laws of complete and incomplete dominance of inherited characters also explained the segregation of dominant and recessive characters in the progeny as per Mendelian ratios. This made it possible to develop homozygous populations of varieties of plant and animal species by crossing parents with differing traits. This revolutionized breeding of plant and animal varieties with high productive yields and desired characters.

The way some of the high input sensitive varieties of plant and animal species are being developed presently raises several concerns regarding environment and sustainability of their production systems. Such varieties though recognized in plant and animal breeding, have no taxonomic identity and require constant monitoring for the selected traits or else they slip into the larger gene pool of species populations.

Despite significant efforts of plant breeding to improve yield of several crops like legumes, cereals and fiber crops, the difference between the potential yield and on farm yields world over have been enormous. Saxena (1990) has documented such differences for the chickpea crop in Syria. The obvious reasons are lack of adequate resistance to pests and diseases, inadequate tolerance to abiotic and biotic stresses, drought, poor nodulation and inadequate nitrogen fixation in legume crops, pollination constraints in crucifer crops etc. However, resistance towards several of these traits exists in wild relatives of several of such crops. Dar (2004) pointed out that high level of resistance to *Heliothis*/*Helicoverpa* damage is available in wild relatives of some crops which can be used in crop improvement through wide hybridization. He stated that genetic transformation and marker-assisted selections have enabled transfer of genes from same and unrelated species which hold great promise in pest management. It is suggested that for better sustainability, the genes for beneficial traits in crop improvement may be selected from larger gene pool that may include sub species or even closely related species and genera.

4.1.4 Biological Species: The only Real Taxonomic Category

The general acceptance of Mayr's (1963) definition of biological species as interbreeding populations reproductively isolated from other such populations by species specific isolating mechanisms laid strong foundations for a debate on the processes involved in biological speciation. Such morphologically distinct, and to some extent reproductively isolated species populations occur only in local habitats. As species are continually evolving dynamic units in response to variable selection pressures in time and geographic space towards periphery of species distribution, even in border areas when species get temporally isolated, processes like hybridization, mutations, polyploidy etc. aided by niche specializations lead to formation of sub species, incipient species and even new species. This does not diminish the relevance of species as the only real category that deals with populations.

During the last 3 centuries, we have discovered only about 2% of the estimated species that are believed to exist. Probably we will never know the exact number of living species. Therefore, with the huge database that now exists at the species level, it is important to know how existing species are related to one another and how they have evolved. Constructing phylogenies is also equally important.

Often widely distributed polyphagous Holarctic species, as *Phytomyza atricornis* Meigen, (Agromyzidae: Diptera), the well known pea leaf miner in India, when examined on the basis of genitalia and pupal characters (Frick, 1959; Griffiths, 1967,1974; Spencer, 1973) was found to consist of two species, the Asian and European populations, *Chromatomyia horticola* (Goureau) feeding on compositae and non-compositae hosts and the American population, *C. syngnesiae* (Hardy) almost restricted to compositae hosts. Species in the genus *Chromatomyia* were earlier placed in the genus *Phytomyza* s.l.

Carolus Linnaeus in his famous work, Xth Edition of *Systema Naturae* published in 1758 laid the foundations of Binomial System of Nomenclature. Large number of typological species of plants and animals discovered and described by him were accepted later as biological species.

4.1.5 Speciation

Multiplication of species or speciation is a constantly evolving dynamic process. Species populations change or enlarge their gene pool through hybridization with differing populations, even with sub-species and related genera, or form new protected gene pools, aided by natural selection. Morphologically distinct geographical isolates of species populations, which are not reproductively isolated, were recognized as sub-species, the only category below the species rank to have taxonomic recognition (Mayr, 1963). In widely distributed plant species, like in conifers, geographic populations showing gradient of characters are often referred to as *clines*. Differing altitude populations showing character gradient are also referred to as *clines*.

Formation of geographical isolates and their becoming reproductively isolated from the main species population is widely accepted as allopatric speciation. Formation of ecological isolates like host specificity, temporal isolates, salt and fresh water isolates of marine coastal populations etc. and their becoming reproductively isolated even in sympatric habitats also became accepted modes of ecological or sympatric speciation.

Rapid Malthusian increase in species populations also occurred many times, when available food resources increased rapidly. The first such expansion occurred in late Cambrian when green algae and other marine plant flora expanded to provide unlimited food resource to several groups of marine invertebrate fauna. It happened again when arthropods and other groups of animals invaded the terrestrial environments during Silurian and Devonian in late Paleozoic. Rapid expansion of dinosaurs and reptile groups in Jurassic and Triassic was also attributed to expansion of Trachaeophytes in terrestrial flora. During adaptive radiation, natural selection eliminates unsuccessful diversification attempts by successive populations. If a group speciates actively, without much adaptive radiation, it will have many species per genus. If any group radiates actively, develops strong divergence, without much speciation, it will have many monotypic genera. Simpson and Beck (1965) and Mayr (1963) have cited numerous examples of molluscs, birds, bats and mammals, which failed to diversify.

It can thus be concluded that speciation and diversification of phylogenies is a dynamic process, having passed through several phases of expansion and extinctions during its geological history. Evolutionary processes guided by natural selection, besides the carrying capacity of environment determine the rate of speciation and the present status of biodiversity.

4.2 Biodiversity: Living Natural Resource

Understanding the true dimensions of living biodiversity natural resources presently seems beyond human comprehension. Despite focussed human effort of last three centuries, we could explore just about 2% of estimated living natural biodiversity resource. This however, does not diminish the relevance of continued efforts to explore this valuable natural resource. Just imagine how much food, nutrition and health security we enjoy from this resource. Also, imagine how life would be if we knew just even 10% of this natural resource. The following section depicts the extent of our present understanding of this natural resource.

Table 1. *Approximate number of known living biodiversity resources*

World Biodiversity	
Vertebrates	45000 including Man
Protists and Invertebrates	60000
Insects	1000000
Lower Plants	100000
Flowering Plants	240000

4.2.1 World Biodiversity Hot Spots

Global hot spots of biodiversity are those regions that contain an exceptional concentration of endemism and experience high rates of habitat loss. The concept of biodiversity hot spots was first developed by Myers (1988, 1990) to identify regions that have exceptional levels of endemism and serious habitat loss. To qualify as a hotspot, a region must have at least 1,500 vascular plants as endemic and must have a high percentage of plant life found nowhere else on the planet. A hotspot, in other words, is irreplaceable. It must have 30% or less of its original natural vegetation, *ie.*, it must be threatened. Around the world, 35 areas qualify as hotspots (Fig. 9). They represent just 2.3% of Earth's land surface, but

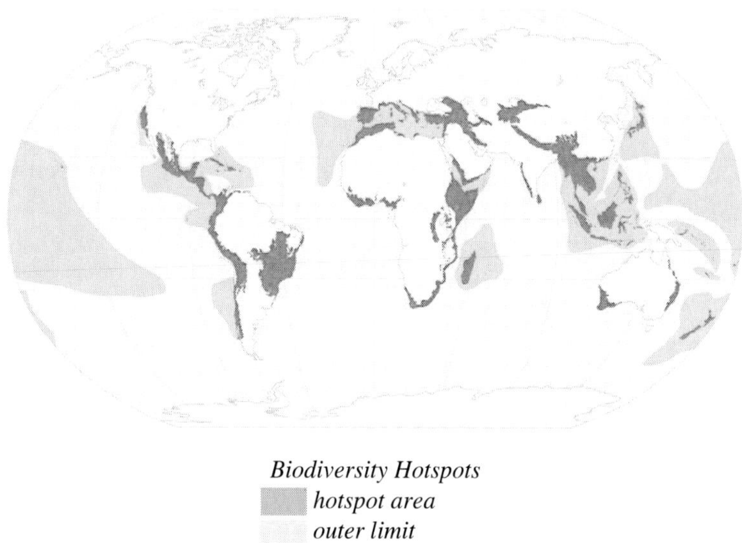

Fig. 9. *Hotspots of world Biodiversity*

they support more than half of the world's plant species as endemics and nearly 43% of bird, mammal, reptile and amphibian species as endemics. Overall, hotspots have lost around 86% of their original habitat and are significantly threatened by extinctions induced by climate change (Malcolm et al., 2006). Without species, there would be no air to breathe, no food to eat, no water to drink. There would be no human society at all. To save the left over, Conservation International (CI) has set up global field offices from Indonesia to Madagascar, Brazil to south-east Asia. These offices located in or near biodiversity hotspots work to protect these places for the benefit of people around the world. CI is an investor in the Critical Ecosystem Partnership Fund (CEPF) which is an alliance of leading conservation donors that provide grants to nonprofit and private-sector organizations working to protect the biodiversity hotspots and improve human well-being. To explore the world's 35 hotspots, access GIS data and learn more about what CEPF and partners are doing to protect these vital places, visit CEPF's website (Conservation International, Biodiversity Hotspots website).

Convention on International Trade in Endangered Species of Wild Fauna and Flora (CITES) adopted resolutions at its convention at Doha, Qatar in 2010 to deal with world issues relating to International Trade in Endangered Species of Wild Fauna and Flora. When a natural habitat, such as a forest is altered to an extent that it no longer supports the species it originally sustained, plant and animal populations are destroyed or displaced, leading to a loss of biodiversity.

Sodhi and Ehrlich (2010) in their introduction to the book *Conservation Biology for All* have stated that our (human) actions have put all humanity into a deep environmental crisis. We have destroyed, degraded and polluted Earth's natural habitat to an extent that majority of populations and species of plants and animals which are part of human life support systems are in decline and many are already extinct. Increasing human population size and consumption per person have precipitated an extinction crisis, the "*sixth mass extinction*", which is comparable to past extinction events such as the Cretaceous-Tertiary mass extinction 65 million years ago that wiped out all the dinosaurs except for the birds. Earlier extinctions were attributed to natural catastrophes like volcanic eruptions, meteorite impact and global cooling; the current mass extinction is exclusively due to deliberate human actions. Being the dominant species on Earth, humans have a moral obligation to ensure conservation of left over rain forests, coral reefs, and tidal pools as well as saguaro cacti, baobab trees, tigers, rhinos, pandas, birds of paradise, butterflies, and a plethora of other creatures. These also provide us with countless invaluable goods and crucial services like food, medicines, pollination, pest control and flood protection. Sodhi and Ehrlich (2010) have further stated that such habitat loss and pollution are particularly acute in developing countries, which is of special concern because these harbor the greatest species diversity and are the richest centers of endemism. Lack of adequate awareness is the main reason that results in major habitat loss. Their observations support the hypothesis that Shrinking Biodiversity leads to habitat destruction which is the prime cause of species extinctions.

4.2.2 World Biodiversity

Biodiversity, the number of living species of plants and animals at any time, is the manifestation of long and gradual process of evolutionary change spanning millions of years in geological time and space. As individuals (humans), we are only privileged to walk a few steps through this beautiful and bountiful heritage of nature.

The present day living biodiversity is the age of flowering plants, the angiosperms, and insects, which are the most dominant and successful members of the terrestrial flora and fauna. These provide almost 90% of food that sustain the rest of natural biodiversity resource including humans. These and the rest of herbivore fauna have coevolved through reciprocal or sequential evolution ever since early Cretaceous almost 200 million years ago.

Marine biodiversity consisting of green algae, phytoplanktons and coastal mangrove flora have supported a rich diversity of consumer organisms consisting of sponges, jelly fishes, complex communities of corals, annelids, arthropods, molluscs, echinoderms, fishes and amphibians over much longer period of 600 million years since Paleozoic era. Today, these resources provide food, nutrition and livelihood to almost 40% of human resource that inhabit along the coastal areas.

Biodiversity is now facing high extinction rates as those during five earlier mass extinctions (Pimm et al., 1995). While reasons for mass extinctions are largely anthropomorphic as destruction of habitat etc., Brito et al. (2012) stated that several natural diseases have the potential to wipe out many species of vulnerable animals. They have developed global hot spot maps of disease distributions where disruptive disease outbreaks could drive several species to extinction. Paul Ehrlich, an eminent ecologist and Senior Fellow at Stanford Woods Institute of Environment has shown that even with extremely conservative estimates, species are disappearing up to 100 times faster than the normal rate (HT 21 June, 2015). This study has warned that we are entering the sixth great mass extinction event and humans could be among the species lost. There is general agreement that extinctions have reached unparalleled level since the dinosaurs died 66 million years ago. If it happens, life would take several millions of years to recover.

During the past few decades, we have lost more natural forests than the present forest cover due to forest fires, systematic clearance of natural forests for farmland and urban needs. Man is now living far beyond the biological means of rejuvenation of lost natural biodiversity. Estimated rate of loss of natural biodiversity resources as stated above is almost 100 times more than the normal rate of species displacement. If unchecked, this together with next phase of mass extinctions due to natural or man made disasters of nuclear wars or even with arrival of more intelligent aliens from cosmos may bring exit of humans sooner than what would normally happen. UNEP (2010) recommends that the Convention for Biological Diversity needs to be strengthened to optimize the benefits that biodiversity conservation can have for climate change mitigation and adaptation by society.

United Nations established the Intergovernmental Platform on Biodiversity and Ecosystem Services (IPBES), amidst growing concerns over the loss of biodiversity and the goods and services it provides to humankind (Brooks et al., 2016). Visconti et al. (2015) tested two widely used indicators of population abundance: geometric mean abundance, equivalent to the living plant index and extinction risk under different climate and land use change scenarios. They tested these on terrestrial carnivore and ungulate species of the orders Carnivora, Cetartiodactyla, Perissodactyla and Proboscidea for which the geographic range was known and available from IUCN (IUCN, 2012), and sufficiently large to obtain an adequate sample of presence points for fitting bioclimatic envelope models. In total, they projected the responses to climate and land-use change impact of 440 of the 493 species in these orders for which range data were available. Their analysis show that both indicators decline steadily, and by 2050, under a Business-as-usual (BAU) scenario, the geometric mean population abundance declines by 18–35% while extinction risk increases for 8–23% of the species, depending on assumptions about species responses to climate change. Based on their analysis, Visconti et al. (2015) observed that aggressive policies to eradicate hunger, ensure universal health, and access to modern energy could be compatible with short-term biodiversity goals. This would require implementation of sustainable production practices, like adoption of reduced logging and sustainable agricultural intensification to increase crop yields. It will also require changes in consumption: low-energy lifestyle, reduce waste and consumption of meat from industrially farmed animals. Environmental legislation, *i.e.*, carbon taxation and strategic placement of protected areas where habitat loss poses the highest threat to biodiversity may be necessary.

This analysis by Visconti et al. (2015) based on carnivore and ungulate groups where distribution and abundance patterns have been well documented indeed provide valuable insight to check biodiversity loss, if sustainable production, consumption and environmental practices are followed. This is not easy. As members of each species population live under constant selection pressures due to competition, predation, multi-dimensional food web, they survive as part of a complex and evolving ecosystem. Their net reproductive rates could vary. Besides, manmade and natural calamities, disease epidemics could also impact populations. Hunting could also be a major threat to species populations (Hoffmann et al., 2011).

Estimates of the number of known species of living plants and animals by pioneering ecologists, E.O. Wilson (1988, 1992) and Robert May (1988) place it around 1.4 million species. Estimates of total number of living biological species range from 3 – 70 million, which suggests that presently we know just about 2% of the total species (Groombridge, 1992; Kormondy, 1996; Raven and Wilson, 1992). Despite this awareness, almost 98% of the total living species biodiversity still remains to be explored. Together with systematic destruction of known species habitats, we have already caused irretrievable damage to life and life support systems.

Table 2. *Number of described species of living organisms and Plants (Wallace, 1876; cf. Kormondy, 1996)*

Common Name	No. of Described species	Total
Viruses	1000	1000
Bacteria	3000	
Blue – green algae	1700	4760
Zygomycete Fungi	665	
Cup Fungi	26650	
Basidomycete Fungi	16000	
Water molds	580	
Chytrids	575	
Cellular slime molds	13	
Plasmodial slime molds	500	46983
Green Algae	7000	
Brown Algae	1500	
Red Algae	4000	
Chrysophyte Algae	12500	
Dinoflagellates	1100	
Euglnoids	800	26900
Mosses, liverworths	16600	
Psilopsids	9	
Lycophytes	1275	
Horsetails	15	
Ferns	10000	27899
Gymnosperms	529	529
Dicots	170000	
Monocots	50000	248428
Protozoans	30800	30800
Invertebrates		
Sponges	5000	

Jelly Fishes, Corals	9000	
Flat worms	12200	
Roundworms	12000	
Annelids	12000	
Mollusks	50000	
Echinoderms	6100	
Insects	751000	
Other Arthropods	123161	
Other Phyla	9300	989761
Chordata		
Tunicates	1250	
Acorn worms	23	
Vertebrata		
Jawless Fishes	63	
Cartilaginous fishes	843	
Bony fishes	18150	
Amphibians	4184	
Reptiles	6300	
Birds	9040	
Mammals	4000	43853
		1392485

4.2.3 Geographical Distribution and Abundance of Species

The distribution and abundance of species on earth is the result of a long history of evolution of phylogenetic lineages globally. As sustainability of human life, biodiversity, agriculture and forestry depends almost entirely on application of biological science to life systems, it is necessary to review them briefly as per faunal regions of Wallace (1876). Closely related species or sister species share a similar geographical space. The major regions are Holarctic consisting of longest global stretch of Nearctic region comprising north America, the Palearctic region comprising much of Europe and Eurasian region and the Oriental region comprising Asia extending to Japan (Fig. 10). Other regions are Nearctic which include north America; Oriental region including India, far east Asia extending to Japan; Neotropical, the south Americas; Ethiopian region includes south Africa and the Australian region. Most of major groups of plants and animal species were fully evolved, before the continental drift started during Mesozoic. There is general agreement that abundance and species diversity increase with decreasing latitudes.

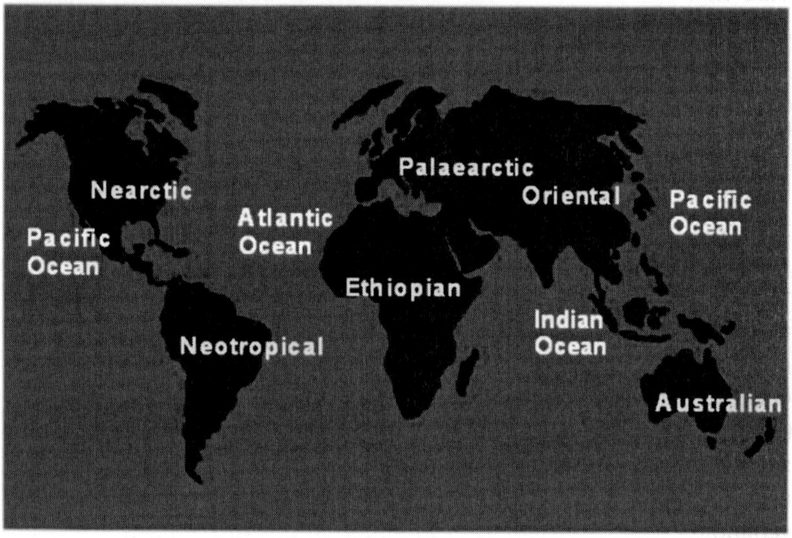

Fig. 10. *Major Biogeographical Regions of the World*

High proportions of terrestrial and fresh water species occur in the tropics. Moving from high to low latitudes, the average species richness within a sampling area of a given size increases as has been documented for a wide spectrum of taxonomic groups (including groups as different as protists, trees, ants, woodpeckers and primates) for data across a range of spatial resolutions (Stevens, 1989; Gaston, 1996).

Such gradients in species richness may be steep (for a given area, tropical assemblages are often several times more speciose than temperate ones), and have been a persistent feature of the species richness, but some debate continues to surround evidence for shallow-water systems, particularly for non-calcareous taxa (Crame and Clarke, 1997). Review by Gaston (2000) has noted that the observed patterns of latitudinal gradients may be at variance. It seems that decline in species richness with latitude may be faster in the Northern than the Southern Hemisphere

(Platnick, 1991; Blackburn and Gaston, 1996). The peaks for richness may not lie at equator itself but some distance away (Roy et al., 1998; Lyons and Willig, 1999). As various groups of biodiversity remain poorly explored, the observed patterns may vary again.

An understanding of spatial heterogeneity of species richness is necessary to understand global distribution of biodiversity. This has vital linkages to the role of biodiversity in ecosystem development, role of invasive species in displacement of local flora and fauna etc. A substantial proportion of regional variation in species richness can be explained statistically in terms of a few environmental variables (Gaston and Williams, 1996; Gaston, 2000). This is, however, far from a predictive theory of species richness. Gaston (2000) believes that for an understanding of expression of patterns of biodiversity, their significance is important to developing such a theory. These processes are affected by ecological factors like the birth, death and immigration and emigration rates of species in an area, which in turn are determined by the effects of abiotic and biotic factors acting at local and regional scales. Rhode (1992) observed relationships between species richness and environmental energy associated with latitudinal, elevational and depth gradients. Gaston (2000) believes that if this were the whole story, patterns in richness would seem reasonably straight forward, if not easy to predict. Such patterns are also influenced by their historical dynamics. These have been shaped by variations in the distribution of different groups of organisms, in their diversification, and hence the availability of species with different attributes to exploit opportunities provided by prevailing conditions. Gaston (2000) suggested that the latitudinal gradient in species richness, however complex it might be, is a consequence of systematic spatial variation in the balance of speciation and the immigration of species, which add species to an area, and of the extinction and emigration of species, which take them away. For very large areas, the effects of speciation and regional or global extinction will predominate, and immigration and emigration will be less important. Blackburn and Gaston (1996) suggested as to why the tropics are highly speciose? According to them, it is most likely due to chance, historical perturbation, environmental stability, habitat heterogeneity, productivity and interspecific interactions. Gaston (2000) further observed that the one factor thought to be important in modulating any effect of the physical structure of the Earth in determining latitudinal gradients in species richness is the relationship between the number of species in an area and ambient available ('usable') environmental energy.

Generalizations like these are difficult to sustain as earth's environment is never uniform over large stretches due to variations in biodiversity, land use patterns, altitude, proximity to tropics as well as the sea coast etc. At best approximations, decline in species richness with increasing latitudes do prevail. The gradient of decline may vary in different groups of biodiversity. However, it seems that intraspecific variations increase due to heterogeneous environment, altitude etc. Such environments may be conducive to formation of geographical isolates; development of host-plant specificities and if these become widespread over isolated areas may give rise to subspecies.

Gaston (2000) concluded that the major terrestrial and freshwater groups are more speciose in tropical than temperate regions, at low elevations than at high, and in forests than in deserts. One might therefore expect that the regional richness of different groups of organisms would co-vary positively and, because of the positive relationship between local and regional richness, local richness would do likewise. This would be important because it would simplify the development of an understanding of global patterns in biodiversity. He has also cited mismatches between the spatial occurrences of peaks in the richness of different groups. For example, among trees, tiger beetles, amphibians, reptiles, birds and mammals, 5% of land area across the United States and southern Canada in which the highest levels of species richness are attained overlap between some pairs of taxa, but this pattern is not a general one. Likewise, although the number of species in different, large and similar-sized areas for two groups are often significantly correlated and may enable a very general impression of the patterns in richness of one group to be obtained from those of another, these correlations are often weak and of rather limited predictive value (Flather et al., 1997). Gaston (2000) observed that in the extreme, some groups exhibit patterns of biodiversity that are entirely contrary to the norm. For example, several major taxonomic groups exhibit peaks of species richness at high or mid-latitudes, (for example, aphids, sawflies, ichneumonids, braconids, bees, and various groups of freshwater invertebrates, marine amphipods, and procellariiformes (Gaston and Williams, 1996; Chown et al., 1998). These exceptions to patterns of biodiversity tend to be observed more frequently at lower taxonomic levels than at higher levels. Distribution and abundance of such groups are probably linked to the favorable ecological environments.

Tropical rain forests cover only about 7% of land surface, yet they contain more than half of the known biological species (cf. Kormondy,

1996). Rapid deforestation and clearing of such tropical rain forests for agriculture, urban needs and forest fires, have put them at irretrievable risk. Beside nutrient poor acidic soils, oxisols, being rich in iron oxide become hard and do not retain calcium, potassium, and phosphorous and other nutrients. It makes regeneration of forests difficult. Kormondy (1996) has estimated that if such clearing of rain forests continues even by 1% every year, all tropical rain forests will be gone by 2035, causing destruction of this largest pool of world biodiversity.

4.2.3.1 Biodiversity: Spatial Distribution (Biodiversity in Americas, Europe, Middle East and Africa; Nepal and South-East Asia; Australia; China and Japan)

The presence of major groups of fauna and flora in different continents indicate that such groups evolved much before the continental drift during late Mesozoic. Differences observed in patterns of geographic distributions of major groups like angiosperms, insects, vertebrates including mammals and homonids are largely attributable to varied selection pressures in different regions after the continental drift. Disjunct populations of different species indicate existence of land connections, or migration routes that may no longer exist today or their extinctions in some areas.

Loss of habitat due to deforestation, forest fires, ecologically unsustainable human settlements, has already put several large animals like elephants, rhinoceros, lions, tigers and unmistakably several of plant species and insects in the list of endangered species. As the theme of this presentation is sustainability of climate, agriculture and man due to shrinking biodiversity during recent times, only a brief account of distribution patterns of different groups of fauna and flora in different regions is given in the following section.

Biodiversity in Americas

As Americas being part of single super continent Panagea were the first to split during continental drift, it does not have several of the distinctive fauna and flora of Africa, Asia and Australia.

The conspicuous absence of several ungulates like elephants, giraffe etc. is a testimony to this. The classic example of the absence of camels in North America is important. Camels originated in North America, from where they spread to south America and Asia. Later they became extinct in North America and Canada due to the last glaciation. After glaciation, North American fauna and flora consisted of elements that moved northwards from South America. It also had significant presence of Palearctic flora that moved in from Europe. Grand prairies of Canada and North America's flat lands form a distinct biome. This has been discussed separately, under the title Grasslands and Mountain Slopes. As stated before, much of this grassland has been transformed into agricultural land and forms the grain bowl of the region.

Biodiversity in Europe, Middle East and Africas

Biodiversity in Europe is the longest evolved flora and fauna adapted to cold temperate sub arctic environments. This has been a contiguous part of the Holarctic flora and fauna.

Much of the Palearctic flora having survived the latest Pleistocene glaciation had conserved the cold temperate flora for the longest spell. Later, this flora spread to the Holarctic region to a large extent. This flora re-established itself in Canada and North American sub-arctic regions after last glaciation and became contiguous with the evolving high altitude flora upto the Himalayas and beyond. As the east coast of South Americas was once joined to west coast of Africa, it shares several of the highly evolved groups of Angiosperms and vertebrate and mammalian fauna of these regions. Flora and fauna of middle-east and South Africa is tropical and sub-tropical that has much in common with that in South America and South East Asia. After continental drift, this largely late Cretaceous flora developed high degree of endemism.

Biodiversity in India, Nepal and South-east Asia

The rich biodiversity in this region is almost entirely attributable to the formation and dominance of Indian Himalayas. These mountain ranges are formed by continuous striking of a subducting of the Indian plate with the Eurasian plate during late cretaceous. As a result, the cretaceous soils from the ocean floor were raised to towering heights to form the Himalayas. As such, several of the sedimentary rocks contain fossil remains of late cretaceous flora and fauna. The northern mountain ranges in India, Pakistan and Nepal merging with the Chinese Mongolian ranges are dominated with Holarctic, Palearctic, Mongolian and north Indian flora and fauna (Fig. 11). Several of these species have evolved into new endemic species.

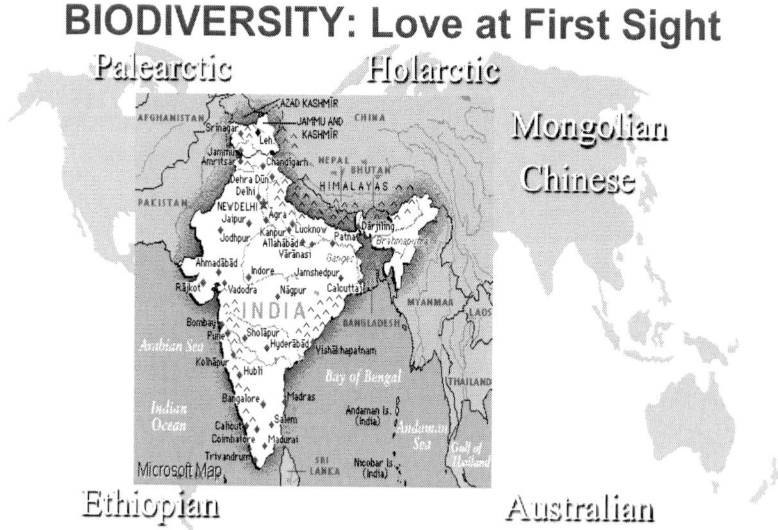

Fig. 11. *Biodiversity Relationships of India, Pakistan and south-east Asia*

India has an extremely rich resource (>8%) of known world biodiversity inhabiting such diverse habitats as coastal plains, rain forests, sub tropical semi-arid environments, deserts, northern plains, low mountain regions together with alpine ranges in Indian Himalayas.

Dr. J.B.S. Alfred, former Director of the Zoological Survey of India, in his meticulous compilation of known faunal diversity in India has estimated about 1,27,000 species of plants, animals and micro-organisms that have been reported so far (Alfred et al., 1998). Amongst these animal species including protozoa, about 89,500 species comprise a major share (7.28%) of total world animal species. According to him, almost 400,000 species of animals, most of which are invertebrates and micro-organisms are yet to explored and identified. The Ministry of Environment and Forests, India, has set up Environment Information System on Faunal Biodiversity at the Zoological Survey of India, Calcutta to computerise and disseminate all available information. This effort alone is unlikely to unravel the enormous biodiversity resources, unless concerted efforts are made at major universities, including institutes of agricultural, forestry and environmental sciences to explore, study, inventory and help conserve such vast biodiversity resources. This is not an easy task in view of the neglect and poor resource allocation to disciplines of taxonomy and systematics, both of plants and animal species diversity. Dedicated human resource has already lost a generation time to undertake this task. Setting up institutional network, training of human resource is an enormous responsibility before the nation, if it must save the glorious heritage of a fast shrinking biodiversity.

Biodiversity in Australia and New Zealand

Australia and New Zealand being part of Gondwana land, share a major part of Ethiopian and south Indian flora and fauna. Kangaroos in Australia, like giraffe in Africa have evolved due to niche specific selection pressures and represent a later and relatively recent diversification. Thus, the flora and fauna is largely semi-tropical and arid, due to large deserts in the regions. New Zealand, which is now believed to be a part of a separate continent Zealandia (Mortimer et al., 2017) has a unique bird Kiwi, which is also its National bird. New Zealand is also famous for Kiwi fruit and has high degree of endemic flora and fauna.

Biodiversity in China and Japan

Much of the biodiversity in China and Japan is adapted to cold temperate region, but also has several elements of north Indian tropical flora and fauna. A lot of Palearctic flora also extends into this region, which mixes with Mongolian flora in eastern regions. Himalayas in this region present a barrier to northward spread of monsoon rains into China. High degree of endemism in this region with sister species in Palearctic and Mongolian region still need to be explored.

4.2.4 Exploring Biodiversity

Despite unequivocal commitment of 150 countries at the historic Earth Summit held at Rio de Janerio in 1992 to promote awareness, conserve forests and biodiversity, nothing much has happened on ground.

Why have we stopped exploration of vast reserves of unexplored biodiversity resources? Hidden under this exploration lies large unexplored gene pools as well as DNA diversity of precious biodiversity resources for welfare and sustainability of humankind. Europe is the best explored region, followed by the Nearctic region. Asia, including China and India, Australia and Ethiopian regions has been relatively poorly explored. Beetles and butterflies, despite being well studied probably still have large unexplored faunas. Similarly, marine fauna and flora are still inadequately explored. Terrestrial tracheophytes and vascular plants are relatively better explored than other groups of plants.

4.2.5 Biological Systematics

Organizing huge diversity of millions of living and fossil species of plants and animals into taxonomic categories, and classifications based on hierarchy of such categories laid the foundations of Biological Systematics. Dr. George E. Ball, an eminent biological systematist and world authority on Carabidae, Coleoptera, defined Biological Systematics as *the description and classification of varied manifestations of uniqueness that characterize the totality of life* (Ball, 1979). He further stated that speciation and phylogeny concern the processes involved in the formation of taxa which are arranged by classifiers in a manner consistent with inferred phylogeny. The species being the basic unit of classification will continue to interest taxonomists and biologists in all groups of biodiversity.

This simplification made it possible to study relationships between species and construct phylogenies based on their phenetic and phylogenetic attributes. This also led to the development of Numerical Taxonomy (Sneth and Sokal, 1973) and Phylogenetic Systematics (Hennig, 1950). Both schools of thought claimed that their classifications reflect true phylogenetic relationships as well. Most taxonomists now use a mixed approach - they define taxa by as many characters as possible.

Organizing enormous available data on species diversity into sets reflecting phylogenetic relationships, their evolutionary history etc. has become necessary to interpret development of modern faunas (Whitehead, 1979) as biological species are no longer considered as strictly reproductively isolated populations in the strict sense as in Mayr (1963). It is only in local faunas, the term biological species as a taxonomically recognizable unit is used. Niche specificity and genotype environment interactions in local faunas complement the morphological and genetic distinction of biological species. As we approach the periphery of species distribution, hybridizations between variable populations become common, leading to the formation of sub species or incipient species. As such, biological species are continually evolving units in response to the variability in selection pressures. Biological systematics has major responsibility to interpret relationships between evolving species populations. A similar question as to why closely related species, which face similar selection pressure show marked interspecific divergence was answered by an eminent Carabid entomologist, Spence, as being due to niche partitioning (Spence, 1979). Besides above, vast amount of data in disciplines like agricultural entomology has accumulated ascribable evidence to geographic or temporally variable species populations, which makes it difficult to interpret results in a meaningful way. The importance of systematics to biological control and Bio-Security has recently been stressed by Malipatil (2001).

4.3 Biodiversification: (Insects, flowering plants and man)

To appreciate the enormous diversity of living biological resources, it is necessary to understand their evolutionary history and kinds of selection pressures that brought this change. Another important lesson from bio-diversification spanning over 600 million years is that evolving geophysical changes had brought with it several phases of mass extinctions and of expansions of biodiversity's resources. Biodiversity has passed through several seral stages of ecological succession where community changes and their replacement by further evolved and probably more successful communities occur with regular frequencies. Despite all this, several major groups of plant and animal species, which once enjoyed a dominant status, have left behind a legacy of few species, which we call living fossils. We still have few representatives of Marcantiales, Filicales, Cycadales, Ginkoales and more in plant world and several species of Xiphosura, the *Limulus*, barnacles, several insects like Thysanurans, Proturans, Collemboles, even cockroaches, lace wings and Palaeopteran orders like Odonata, Ephemeroptera and more in the animal world which still survive and occupy specialized ecological niches of their own. Such species form part of our glorious heritage and give a glimpse of the past natural history of such groups.

As our prime concern is sustainability of man and his agriculture, and of major groups of biodiversity resources, only selected groups of such resources are discussed below. Flowering plants, the Angiosperms and vastly diversified insect fauna have coevolved through reciprocal or sequential evolution, ever since early Cretaceous, about 200 million years ago into a rich diversity of dominant living biodiversity resources in terrestrial environments. Man, the latest arrival, about 0.1 million years ago, has already set his footprints in every living ecosystem on Earth.

4.3.1 Insects and Flowering Plants

4.3.1.1 Insects

Insects represent the most remarkable success story in bio-diversification. Although the first fossils of wingless insects are known from Devonian about 400 m years ago, it took almost 100 m years until late Carboniferous and during Permian, for insects to evolve functional wings, when most insect orders as we know them became diversified. Evolution of wings must have been a great ordeal for the wingless insects. As larger shrubs and tall trees began to grow on earth, such insects must have climbed tree tops for food and escape from predators. High winds and frequent falls from tree tops must have forced them to be air borne and provided the much needed selection pressure for lighter and flattened bodies and enlargement of their paranotal lobes to act as gliding planes as seen in the fossil Paleodictyoptera. Such primitive insects soon developed wing articulation and diversified into two lineages, the Paleopteran and the Neopteran orders. The Paleopteran orders lacked the wing flexing mechanism and could not survive competition and only the most efficient amongst them, the dragonflies and the mayflies are represented by two living orders, the Odonata and the Ephemeroptera. The more successful Neopteran lineage, soon gave rise to orthopteroid orders, where young ones almost resembled the adults. The most generalized amongst them retained the primitive hypognathous orientation of the head capsule convenient for feeding on open vegetation while the more specialized developed the prognathous orientation for chasing prey or food. Some members of this stock, much like the baby elephant had their mouth parts pulled up to form the proboscis to reach the hidden foods beneath the plant and animal surfaces and gave rise to the hemipteroid orders. While some other members of the orthopteroid stock gave rise to holometabolous endopterygote orders, where the young ones did not look like the adults and after a voracious meal required a resting or pupal stage to fully transform into the winged adult. One lineage amongst them gave rise to the related orders in the Panorpoid complex of Mecoptera, Neuroptera, Siphonaptera, Trichoptera, Lepidoptera and Diptera; while the other gave rise to most successful orders of Coleoptera and Hymenoptera. Almost 40% living insects are beetles. Why are there so many beetles? According to Farrell (1998), repeated origin of angiosperm-feeding beetle lineages associated with enhanced rates of diversification are responsible for this. Similar logic together with high degree of endemism and poor dispersal ability of most species explains this extra-ordinary diversification of insect species.

4.3.1.2 Flowering plants

Plant life during the Paleozoic consisted largely of Lycopsids, Sphenopsids, Pteropsids, Filicales and the Cycadofilicales. It was most likely the Pteropsid stock that gave rise to the flowering plants. It was only during Mesozoic that maximum diversification of flowering plants occurred. As such, plants were subjected to multiple selection pressures due to feeding by phytophagous insects, mites, other herbivores and also in response to disease pathogens and for adaptation to the abiotic environments, some of those which evolved a complex array of secondary plant chemicals entered a new niche protected from the earlier group of herbivores and other biotic and abiotic stresses and thus were better adapted to diversify. As some insects and other herbivores evolved specific enzymes, the mixed function oxidases, largely as a genetic feedback mechanism, started feeding on earlier protected plants by detoxification of the secondary plant chemicals. This reciprocal evolution continued and it was referred to as *Coevolution* by Ehrlich and Raven (1964) and as *Sequential Evolution* and now as *autonomous evolutionary events* by Jermy (1993). Since Mesozoic, angiosperm flora during their reciprocal evolution evolved not only the most sophisticated natural defences for their own survival, but also evolved the rich genetic diversity of plant foods which provided the basis for modern agriculture. About 10,000 plant metabolites have been documented so far and the estimates of the total number of organic chemicals indicate more than 400,000 chemicals in the plant kingdom (Grainge and Ahmad, 1988). Ever since Cretaceous, 200 m years ago, primarily due to such large diversification of plant chemicals and reciprocal diversity of insect feeding, both flowering plants, the primary producers and insects, the primary consumers and other herbivores lived in complete harmony with each other and the environment.

All ecosystems ultimately reach the stage of self sustainability due to their innate homeostasis supported by Multi Species Life Support Systems (MLSS). There are no pests in such ecosystems. Pests in any evolving ecosystem including forests and agricultural ecosystems are mainly a population problem. The larger the number of species, the lesser will be numbers per species population.

4.3.2 Man: the *Homo sapiens*. Here, There and Everywhere

There is general agreement that the hominid line that gave rise to Man, separated from the Pongid line which comprised three genera; genus *Pan*, the chimpanzee and gorilla in Africa; genus *Pongo*, the oranguton, East Indies Java man; genus *Hylobates*, and gibbon group in South-east Asia. The most striking difference between man and other anthropoids is in brain size which in anthropoids is one-third that of man. During mid-to late Pleistocene, three definite stages in human evolution were established. The *Australopithicus* stage in Africa, characterized by massive skeleton and large teeth; the *Homo erectus* stage in Africa, Europe and Asia and the *Homo sapiens* stage with a relatively larger brain. All these also indicated changes in axial skeleton relevant to a shift from arboreal to bipedal life. The Java man, *Homo erectus*, the Pekin man, *Sinanthropus pekinensis* were close to *Homo sapiens* which in *sensu lato* formed a polytypic species. With subsequent migrations as man set his footprints on every corner of Planet Earth, introgressive hybridisation between differing populations has blurred any distinction of sub-species, even of races and morphotypes. He is now a single monotypic genus, with little possibility of any further speciation.

Has man stopped speciating completely? It may not be so. It has been a long journey from man as animal to man as social human being. This aspect is still evolving as evidenced by medieval barbaric ways of life in fairly large sections of population. Slow and gradual transition took him through Iron age and tribal nomadic ways of community living. They still cultured territoriality and fought inter-tribal wars of dominance. Although, during this period he became relatively more intelligent, learning skills for better adaptations to life suited to hunting and exploiting forest resources for food and shelter. This led to gradual shortening and better use of fore limbs, improved parental care of young ones and above all, further improved learning and brain body ratio. Human behaviour as we now know is almost entirely learned by exposure to immediate life environment.

Man, for too long, has considered himself as the most dominant and successful biological species of the vertebrate lineage. This perceived dominance of man is due to the intelligence acquired through his mostly learned behaviour by which he has adapted the environment and rest of biodiversity to serve his needs. This claim is belittled by another dominant group of terrestrial animals, the insects. By contrast, insect behaviour is largely inflexible and is modified genetically during evolution. As insects are still evolving at a much faster rate than most other animal groups, they are at least twenty times more numerous and more diversely adapted than the vertebrates. A single swarm of locusts could weigh up to two thousand tons, which when computed at 2.5 g per locust, will consist of over eight hundred million locusts. A single beehive may have over fifty thousand bees. Similarly, grasshoppers, termites, armyworms, ants, wasps, parasitoids etc. have equally large numbers. The total world population of insects has been estimated at 10^{18} and even at most conservative estimate of 2.5 mg per insect, the total weight of insects will exceed that of man by more than twelve times (Hocking, 1968). Obviously, insects are the most numerous, if not, dominant members of our most fragile ecosystem, the biosphere. With almost one million known biological species, insects form about three-fourth of the total animal species biodiversity, and almost two-third of the entire organismic species biodiversity, including that of plants and protists. The fact that vast areas of Asia, Africa and Latin America are still poorly explored and the rate at which new species have been discovered in the past, indicate that the actual insect species biodiversity is at least ten times more than what is presently known. The estimates of insect species that have become extinct are equally high.

4.3.2.1 Poverty of Human Resource

What is poverty? Why there are so many poor, deprived people in the most productive part of the world? Although much has been written on this subject, an attempt is made to answer this from within the biological domain. Human needs of food, fibre and shelter do not differ with caste, class or status in a civilized world. Poverty is denial of this inalienable right to vast populations by the ruling political classes of all shades who define different set of rules for their own welfare. Unemployment likewise, is absolute failure of implementation of policies for deployment of the precious human resource in productive pursuits by the state. In thickly populated countries as in India, Asia, parts of Africa and Latin America, where millions are dying due to poor infrastructure and inadequate hygiene, this vital human resource can become the agents of change, only if their basic human needs of food, nutrition, shelter, health and social security are met with equitable sharing of gross national resources in any country. If there is political will and recognition of this social responsibility, irrespective of caste, creed, gender and social status, no price is too big to meet this obligation. It is not enough for United Nations and FAO to mobilise food aid; they should also monitor that food aid reaches the poorest of the poor, where it is needed to prevent starvation and mass famine. This world produces enough food; even more than what is required to prevent hunger and malnutrition in the precious human resource at present levels of world population. This human resource, if deployed with social responsibility in productive pursuits of building infrastructure, management of natural resources and to meet the ever growing demand of upcoming industries, can become an agent for change. Adequate minimum wage to take care of basic needs per family of 4 to provide for their nutrition, health and life support systems including the old age support needs to be a part of carefully planned package of state social welfare.

According to International Assessment for Agricultural Knowledge, Science & Technology for Development, IAASTD (2009) Global Report, *Agriculture at a Crossroads*, over 70% of world's poor live in rural areas. These 2.1 billion people live on less than US$ 2 a day. The livelihood of many such farmers/ farm workers is oriented to meet basic human needs, mainly food. With insufficient income, most households have little money to improve production and sustainability of their production systems. Given greater social equity at local, national and global level, world agriculture has the potential to provide better livelihood, nutrition, sanitation, health care and better life's environment. The IAASTD report further notes that a combination of various factors including poor health, reduced work capacity, low productivity, short life expectancy is typical for people engaged in agriculture. Despite sufficient food calories available, millions die or are debilitated every year by hunger and malnutrition.

Let me state once again that thickly populated countries like India are neither resource poor nor resource deficient to honour this obligation. Imagine, while we are losing 40% of food production between harvest to mouth, over millions of tons of food grains rot along road sides during monsoon rains, due to lack of adequate storage structures and post harvest handling facilities. This is true not only in respect of food grains, it is equally true with fruits, vegetables and several other crops. Employing youth in enhancing, conserving and nourishing biodiversity resources is an ever available strategy.

4.4 SHRINKING BIODIVERSITY: CONCEPT AND IMPACT

The theme of this presentation is based on the perceptions that after 1940's second world war, discovery of DDT and Malthusian increase in human populations triggered by green revolution of early 1960's, exploitation of natural biodiversity resources increased far beyond the nature's ability to renew them. Even the quality of human resource decreased as evidenced by many fold increase in incidence of cardiac, respiratory and cancer diseases due to consumption of pesticide residues in food crops and dairy products.

Global loss of animal species is like deforestation in terms of scale and impact. Though for emotional or aesthetic reasons we may lament the loss of large charismatic species such as tigers, rhinos and pandas, we now know that loss of animals, from the largest elephant to the smallest beetle will also fundamentally alter the form and function of the ecosystems upon which we all depend (Dirzo et al., 2014). Identifying the drivers of these extinctions is straight forward, but stemming the loss is a challenge. Animal species continue to decline and disappear from even large, long-protected reserves due both to direct impacts, such as poaching and indirect ecological feedbacks, such as habitat fragmentation. Though hunting and poaching might seem obvious candidates for targeted policy and management interventions, there are complex social issues underlying these activities that will require coordinated and cooperative actions by nations (Brashares et al., 2014). Attempts to reverse the extinction trend are now increasing. Such efforts involve a variety of approaches, including breeding animals in captivity, with the hope of reintroducing them to the wild, and assisting recolonization of areas where species have become locally extinct (Seddon et al.,2016). Active reversal of animal extinctions is proving just as challenging as preventing extinctions in the first place, but a few success stories provide some hope. Many note and mourn the loss of animals but have not recognized that the impacts of this loss go beyond an aesthetic and emotional need to maintain animals as a part of nature. Current research reveals startling rates of animal declines and extinctions and confirms the importance of these species to ecosystems (Stokstad, 2014). If we are unable to end or reverse the rate of their loss, it will mean more for our own survival than just an empty forest. Interactions between environment and evolving populations, like competition for food, space, predation etc. is the major cause driving many species to extinction. The Shivalik ranges in northern Himalayas up until the early 1970's were abundant in cold temperate Palearctic flora, like *Ranunculus, Thalictrum, Clemetis, Delphinium, Ferns, Plantago, Asters, Solidago* etc. Forest fires, deforestation for urban needs and invasion by aggressive alien species like *Lantana*, *Parthenium* etc. has since made such flora a rare sight. Rapid extinction of dinosaurs and other groups during late Cretaceous occurred due to many reasons, which included onset of northward continental drift, competition, origin of more intelligent forms of vertebrates and primates with better brain body ratios. When continental shelves passed through alternate periods of flooding and relative dryness, there being few geographic barriers, multiplication of species was very slow. Periodic extinctions of marine invertebrates during Devonian, Carboniferous and in Permian are partly attributed to this phenomenon.

Vignieri et al. (1914) emphasized that in the recent epoch Pleistocene, only tens of thousands of years ago, Planet Earth supported large animals like mammoths, terror birds, giant tortoises and saber-toothed cats, as well as many less familiar species such as giant ground sloths (some of which reached 7 meters in height) and glyptodonts (which resembled car-sized armadillos) which roamed freely. Since then, many such species were driven to extinction due to many reasons which include larger diversity of animal species, including the arrival of Primates including humans with better brain body ratio. Animal species are rapidly disappearing in Borneo, Amazon, Europe, South-east Asia and around the world.

4.4.1 How living systems are affected?

Clearing forests, forest fires, industrial pollutants, cutting trees for fuel, unplanned urbanization etc. are often the main causes of shrinking of natural biodiversity resources. This affects every living natural ecosystem, depriving it of multi species life support systems. Life on planet earth exists as inter dependent complex of biological species that co-exists in any given habitat. Such habitats interact with other similarly living habitats eventually leading to the entire biosphere as a composite ecosystem.

Hooper et al. (2005) brought forth the consensus statement how biodiversity affects ecosystem functioning. The significance of biodiversity for welfare of mankind was recognized during the Earth Summit 1992 at Rio de Janeiro, which led to the formation of the Convention on Biological Diversity, endorsed by intergovernmental agreement of 193 countries. This included conservation of biological diversity, the sustainable use of its components, and the fair and equitable sharing of the benefits. Despite this agreement, biodiversity loss continued at increasing rates (Heywood, 1995; Butchart et al., 2010). Without an understanding of linkages between biodiversity, ecosystem functions and the benefits that accrue from them, it is difficult to achieve any policy objectives (Larigauderie et al., 2012). With fundamental knowledge about ecosystem functioning, it may be possible to manage biodiversity loss. Schulze and Mooney (1993) gave several original ideas that led to empirical researches on biodiversity and ecosystem functioning, a concept popularly known as BEF. Loreau et al. (2002) reviewed researches on biodiversity and functioning of the ecosystem. Daily's (1997) concept that natural habitats provide goods and services to society, highlighted ecosystem services.

There is general agreement that human actions are damaging Earth's ecosystems, eliminating genes, species and biological traits at an alarming rate. This biodiversity loss is a major driver of ecosystem change (Hooper et al., 2012). Latta et.al. (2011) documented how species and genotype diversity drive community and ecosystem properties in experimental microcosms.

Meta-analysis by Cardinale et al. (2012) show that reduction in number of genes, species and functional groups of organisms reduce efficiency by which ecological communities capture biologically essential resources, produce biomass, decompose and recycle biologically essential nutrients. They stressed importance of close intellectual ties between biodiversity ecosystem functioning (BEF) and biodiversity ecosystem services (BES). If we want to know how biodiversity influences the ability of the ecosystem to remove carbon dioxide from the atmosphere or store carbon over long time frames, then we need to consider the net influence of biodiversity on photosynthesis and carbon sequestration besides accumulation of carbon in the live plant. They documented that once organisms die, leaf litter diversity enhances decomposition and recycling. They noted that biodiversity effects seem to be remarkably consistent across different groups of organisms, among trophic levels and across the various ecosystems that have been studied. This consistency indicates that there are general underlying principles that dictate how the organization of communities influences their functioning. Their meta-analysis of available researches during last two decades has led to six conclusions:

- Biodiversity loss reduces the efficiency by which ecological communities capture biologically essential resources, produce biomass, decompose and recycle biologically essential nutrients.
- There is mounting evidence that biodiversity increases the stability of ecosystem functions through time.
- The impact of biodiversity on any single ecosystem process is non linear and saturating, such that change accelerate as biodiversity loss increases.
- Diverse communities are more productive because they contain key species that have a large influence on productivity and differences in functional traits among organisms increase total resource capture.
- Loss of diversity across trophic levels has the potential to influence ecosystem functioning even more strongly than diversity loss within the trophic levels.
- Functional traits of organisms have large impacts on the magnitude of ecosystem functions, which give rise to a wide range of plausible impacts of extinction on ecosystem function.

The above conclusions reinforce our perceptions that multi species life support systems and larger biological diversity broadly leads to greater sustainability of production systems. To sum up, there is greater awareness how biodiversity loss reduces the efficiency by which ecological communities capture biologically essential resources, produce biomass, decompose and recycle biologically essential nutrients.

Cardinale et al. (2012) have further examined whether biodiversity loss is sufficiently large enough to rank among major global drivers of environment change. They cited a study by Tilman et al. (2014) which shows that biodiversity impacts ecosystem productivity as much as resources, disturbance or herbivory. This study compared 11 long term experiments performed at one site. Another study by Hooper et al. (2012) which used meta-analysis from published data to show that the impact of species loss on primary productivity are of comparable magnitude to the impacts of drought, ultraviolet radiation, climate warming, ozone, acidification, elevated CO_2, herbivory, fire and certain forms of nutrient pollution. As Biodiversity Ecosystem Functioning (BEF) relationship is non-linear, the exact ranking of diversity relative to other drivers will depend on the magnitude of biodiversity loss and of environmental changes. These two studies indicate that diversity loss may have a significant impact on ecosystem functions as other global change stressors like climate change (Houghton et al., 2007). Cardinale et al. (2012) observed that diversity effects grow stronger with time and may increase at larger spatial scales. Short-term experiments may underestimate the impacts of diversity loss on the functioning of more natural ecosystems (Duffy, 2009). However, at larger spatial scales and with greater temporal fluctuations, more environmental heterogeneity may increase opportunities for species to exploit more niches (Cardinale et al., 2007). Dispersal of species also scales up biodiversity - productivity relationships (Venail et al., 2010). High plant diversity is required to maintain ecosystem services and ecosystem multi-functionality (Hector and Bagchi, 2007; Zavaleta et al., 2010; Isabell, 2011).

Cardinale et al. (2012) have suggested that the ecological consequences of biodiversity loss can be predicted from evolutionary history. Species richness has often been used as a primary measure of biodiversity. Species genepool and variations are determined by patterns of common ancestry (Mace et al., 2003). Phylogenetic distance or genetic divergence among species may explain variations in biomass production than taxonomic diversity (Cadotte et al., 2008; Flynn et al., 2011). This suggests that evolutionary processes that generate trait variation among organisms are, in part, responsible for the ecosystem consequences of biodiversity loss (Cardinale et al., 2012).

The exhaustive review by Cardinale et al. (2012) makes a stimulating reading and has wide applications for sustainable natural ecosystems. Their conclusions that intraspecific genetic diversity increases the yield of commercial crops; tree species diversity enhances production of wood in plantations; plant species diversity in grasslands enhances the production of fodder; and increasing diversity of fish is associated with greater stability of fisheries yields are in conformity with traditional diversified agricultural production systems that farmers with small holdings (>2 ha) follow in developing countries as India. This is what has enabled them to feed a nation of 1.2 b people. Although hunger and poverty still prevail, it is more due to the political system and an inequitable sharing of agricultural productivity. For other regulating processes they concluded that increasing plant biodiversity increases resistance to invasion by exotic plants, plant pathogens such as fungal and viral infections that are less prevalent in more diverse plant communities. Plant species diversity increases above ground carbon sequestration through enhanced biomass production; nutrient mineralization and soil organic matter increases with plant richness and is in consonance with the lessons we learnt from sustainability of diversified agricultural production systems in small farm traditional practices.

Relationship between species richness and ecosystem processes increase linearly with species richness (Johnson et al., 1996). This relationship is likely to saturate as niche overlap increases at higher levels of diversity (Vitousek and Hooper, 1993). Chapin et al. (2000) have concluded that human alteration of the global environment has triggered the sixth major extinction event in the history of life and caused widespread changes in the global distribution of organisms. They have cited several examples from North America in support of their hypothesis. Such alterations, extinctions, and colonization of biodiversity resources have increased several times since last 10,000 years when man took to agriculture. The spate of such changes intensified in tropical and semi-tropical world after 1960's adoption of green revolution crops technologies. Deliberate clearing of understory in thick natural forests have almost eliminated natural flora like *Thalictrum, Delphinium, Clematis, Artemisia, Cassia* etc. from roadsides and from lower altitudes in the foothills of Himalayas. This space was in no time invaded by robust exotic species, *Lantana camara*, *Parthenium* etc.

Ecological restoration of degraded ecosystem is essential to restore its health. United Nation Convention on Biological Diversity has defined it as "*assisting the recovery of an ecosystem that has been degraded, damaged or destroyed*". Jorgensen (2013) suggests that the goal of 15% recovery by 2020 should be modified to allow progress measurement against the numerical target, or the target should be modified to match the chosen definition. Life on planet earth exists as inter-dependent web of biological species that coexists in any given habitat.

4.5 SHRINKING BIODIVERSITY: LIVING ECOSYSTEMS

Life in living ecosystems is sustained by the ultimate source of energy, the solar radiation. During photosynthesis, water from soil and carbon dioxide from air is used in the presence of chlorophyll and sunlight to form glucose, which is transformed into other sugars, cellulose, lignins, proteins and wood. Photosynthetic efficiency in typical green crop plants; i.e., absorption of solar radiation for conversion to plant biomass is 1-2% as below:

$$6H_2O + 6CO_2 + \text{energy} \longrightarrow C_6H_{12}O_6 + 6O_2$$

Planet Earth is home to several kinds of ecosystems, where energy flows by nutrient biocycling from chemical and physical environment to autotrophs/producers, herbivores, carnivores, decomposers which supports inter-dependent web of biological species in any given habitat. As per second law of thermodynamics, some energy is lost as entropy when it passes from one trophic level to another. Loss of energy when herbivores feed on producers as crop plants is minimum (~ 15%), whereas it is maximum (~30%) when carnivore feed on herbivores or carnivores feed on carnivores. Shorter food chains as vegetarian diet for human populations is most efficient than the meat diet.

In any ecosystem, members of different species live as interacting communities of organisms, each creating their own space, their very own intra- and inter-specific communications, their own social care and defense systems etc. which ensures their survivability. Such ecosystems with larger number of species are more stable.

Phenological events in terrestrial flora and fauna like flowering, maturity, arrival of spring, autumn etc. show latitude, longitude and altitude variations as per Hopkin's Bioclimatic Law. These variations are at the rate of four days for each degree of latitude, later north, five degrees of longitude, later east and 400 feet of altitude later upwards (Hopkins, 1919).

The largest ecosystem on earth, the biosphere, consists of two major types - terrestrial and aquatic. Terrestrial ecosystems are of several types as tropical rain forests, deserts, temperate deciduous and rain forests, mountains, temperate grasslands, boreal conifer forests, tundra etc. Aquatic ecosystems consist of freshwater, marine, coastal, estuarine, deep sea etc. Each of these eosystems have suffered biodiversity losses due to natural and man made calamities, periodic extinctions, land use for agriculture and urban needs etc. Shrinking biodiversity resources beyond the nature's ability for their renewal has had adverse impact on sustainability of climate, agriculture and man. Tropical rain forests have maximum biodiversity and biomass. Other major ecosystems from equator to arctics have gradual reduction of biodiversity and biomass. In the following section, some of these major ecosystems are discussed.

4.5.1 Forests

Natural forests are not mere collection of trees. They are evolving dynamic ecosystems to sustain energy flow from producers to the top carnivore through different trophic levels involving a complex chain of food webs. Mature ecosystems sustain equilibrium through innate homeostatic mechanism. FAO data suggests that except East Asia, Caribbean and Western and Central Asia, the average rate of annual changes in Global Forest Cover have been in negative with maximum decline in Central America (> -1.2%), followed by that in Northern Africa, South and Southeast Asia, East and South Africa and South America (Fig. 12). This news release stated that Hansen and colleagues analyzed 143 billion pixels in 654,000 Landsat images to compile maps of forest loss and gain between 2000 and 2012. During that period, 888,000 square miles (2.3 million square kilometers) of forest were lost while 309,000 square miles (800,000 square kilometers) regrew. The researchers, including scientists from the University of Maryland, Google, the State University of New York, Woods Hole Research Center, the U.S. Geological Survey (USGS) and South Dakota State University published their work in Science (Hansen et al., 2013). Key to the project was collaboration with team members from Google Earth Engine, who reproduced in Google Cloud, the models developed at the University of Maryland for processing and characterizing the Landsat data.

This data shows that between 2000 and 2012, maximum losses occurred in Tropical Rain Forests (601,071 sq.km), followed by that of Boreal coniferous forest (350,135 sq.km), Tropical moist deciduous forest (300,149 sq.km), Sub Tropical humid forest, Boreal Mountain Forest, Tropical dry forest and Temperate zone forests (Fig. 13). This study is very inspiring besides suggesting that shrinking natural forests together with shrinking natural biodiversity resources especially in Tropical Rain Forests of Indonesia, and other equatorial states, as also in Central America, even in Canadian Rockies and in 100 kilometer wide range of lowland regions along the entire length of the foothills of Indian Himalayas are the major cause of deterioration of World Climate and unsustainability of World Agriculture. This damage is severe in thickly populated countries of Asia and Latin America, where land forest ratio has significantly declined due to 1960's green revolution high input technologies of chemical fertilizers and chemical pesticides. Such technologies have made many countries self-sufficient in food grains and this led to Malthusian increase in populations. As people in largely unmonitored environments are now consuming high levels of pesticide residue on their dining table, this together with high levels of air pollution and weakened immune system, incidence of diseases like cardiac and respiratory failures, arthritis etc. has increased many fold. Sustainability of food, nutrition, health and social security of almost three-fourths of world populations, in the absence of meaningful State Social Welfare is at stake. Such damage in countries where land to forest ratio is high and are not thickly populated like USA, Canada, parts of Europe, Australia etc., lesser impact on populations has been observed.

Annual Changes in Global Forest Cover

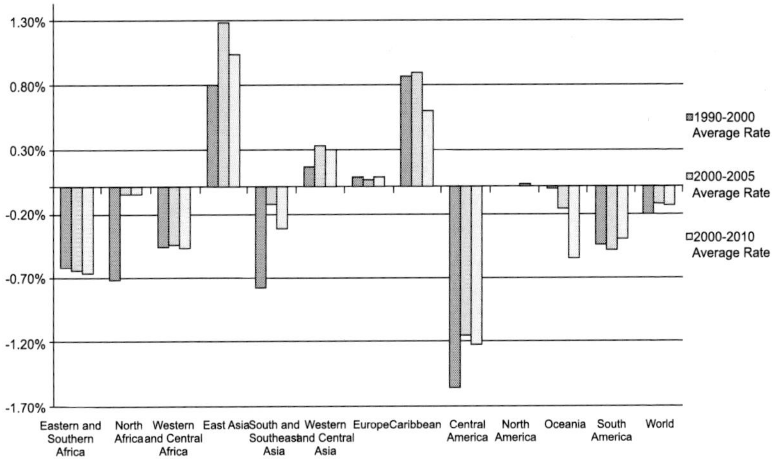

Source: FAO-Global Forest Resources Assessment 2010 at www.fao.org/docrep/013/i1757e/i1757e/i1757e.pdf

Fig. 12. *Annual Change in Global Forest Cover*

Forest loss by ecozone, 2000-2012 (sq km)

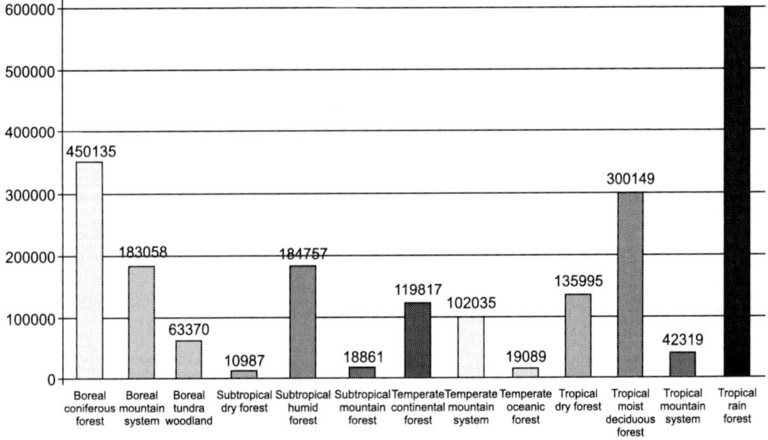

Fig. 13. *Forest Loss by Eco zone.*

Maximum losses of Tropical Rain Forests, 10 degrees north and south of equator, from Central and South America to East Indies and northern Australia are alarming as such forests occupy only 7% of the land surface yet they house more than 50% of world biodiversity (Kormondy, 1996). These forests are among the oldest known ecosystem that existed for more than 60 million years (Richards, 1973). These forests exert significant influence on world climate, global balance of carbon and atmospheric pollution. Being rich in species diversity, they provide new sources of food, fiber and medicinal plants (Jordan, 1985).

Unwanted human intervention leads to destabilization and threatens sustainability of natural ecosystem. If a top carnivore becomes an endangered species, as is with big cats in many parts of the world, thousands of species of animals, plants and microbes also perish. A forest tree takes almost hundred years to grow, whereas human greed can cut it in less than a day. During the last few decades, we have destroyed more forests than the present forest cover.

A report in Newspapers (TOI,15.08.2015) that Uttarakhand Government citing forest fire risks intend to remove several thousands of 'Chir Pines' long needle Pines, *Pinus longifolia,* from lower to mid hills of central Kumaon Himalayas in India. These trees having evolved through 100 million years are an important part of forest ecosystems in the area. The truth is that other valuable trees like, oak, balsam, poplars, and other pine species have been systematically cleared for firewood, timber etc. Instead of rejuvenation of such forests with massive reforestation with other trees, and promoting pear, plum and apricot orchards in such areas, cutting several thousands of pine trees is no solution. Biocycling of unwanted pine needles to produce organic manure could also be considered. The rate at which such forest resources are getting depleted is far beyond the biological means of their renewal.

As forests provide a big share of human need of food, nutrition, medicinal and health supplements, dying forests would be major blow to the very sustainability of man and other biodiversity resources. Forests also absorb large amounts of harmful gases like carbon dioxide, methane, sulphur and carbon monoxide, thus providing us with clean air and oxygen besides protecting the ozone layer for sustainable life's habitat. Shrinking forests have pushed major biodiversity resources to few protected refugias. Such refugias are restricted to national parks and or along northern slopes of mountain ranges like Himalayas, Alps and

Rocky Mountains in America etc. Tropical and semi-tropical rain forests, the richest reserves of biodiversity resources, are also disappearing at a faster rate than ever before. Biennial State of Forest (SFR) Report (2013) from the Ministry of Environment, Forest and Climate Change, India, states that forest cover in India increased by 3.3 million hectares between 1999 and 2013. Much of this increase in forest cover is due to proliferation of plantation crops like tea, coffee, orchards etc., outside forest areas, masking massive deforestation in the natural forests. This disparity came to light from analyses of methodology and data adopted in Forest Survey of India (FSI) reports in the last few years. Due to technical limitations, lack of digitized data, FSI included green patches also known as '*green wash areas*' in the FSI forest cover reports. As per the SFR (2013) report, out of a total of 69.8 million ha forest cover, 53.08 million ha lie in green wash areas. A re-evaluation of forest cover data of 1999 shows that green wash area was 63.66 million ha (SFR, 2009). This shows that between the years 2009-2013, the country has lost 10.58 million ha out of a total of about 69.8 million ha of original, natural forests. Such depletion of original forests in Kashmir, Assam, Himalayas and in Gujarat are in the range of 50-70%; while in central and peninsular India, where much of the rain forests are situated, losses are 0-25% (HT Report 18 July 2015 based on State of Forest Reports 1999; 2009 and 2013; Forest Survey of India (FSI) Report, (2013).

Ever since 1960's, large contiguous tracts of original forests were cleared for agriculture, for adoption of green revolution crops of wheat and rice all along the northern plains of India. Though increases in green cover due to plantations like tea, coffee and orchards are welcome, but real biodiversity reserves lie in original natural forests. Similar losses due to forest clearings have occurred in tropical rain forests worldwide. Such losses are largely due to clearing forest land for agriculture, urban habitats and due to forest fires. Similarly, deforestation of mangroves along the marine coastal habitats along Western Ghats in peninsular India and along the Bay of Bengal has inflicted heavy losses of biodiversity.

4.5.1.1 Empty Forests

Over hunting is one of the major causes of vanishing forest biodiversity, more so in Asia, Africa and South America. A recent case study of Lambir Hills National Park in western Borneo's most diverse natural forest by Stokstad (2014) has revealed that over hunting and illegal trade of wild life has depleted herbivores that thinned saplings and fruit eaters that dispersed the seeds. Thus, crowded trees compete for nutrients and resources. Such trees may soon become sick and the forest also loses species diversity. A total of 239 rhinos were killed in 16 years in Assam forests due to poaching (Shillong Times, 22/07/2016). Similar loss of big animals like elephants, tigers etc. and forest flora is turning our natural forests into a mere collection of trees. Attempts at refaunation to restore biodiversity are relevant, though these have met limited success (Seddon et al., 2016).

4.5.2 Grasslands, mountain slopes, valleys and plains

Together with forests, grassland ecosystems along mountain slopes, valleys and in plains not only bind soils and prevent erosion, but also support human need for food, fodder, fuel and industrial raw material.

Prairies are the temperate grasslands with predominant composition of grasses, herbs and shrubs, which form a distinct biome. These regions are predominantly located in North America and Canada, although similar regions are also found in Argentina, Brazil and Eurasia. The following brief account of the Canadian Prairies has much in common with most of these regions. The great prairies of Canada cover the vast flat lands in the south of Alberta, Saskatchwan and Manitoba provinces which form the richest conserve of North American grasslands, now support a specialized ecozone with distinct flora and fauna. Gas and oil are abundant in this region. Several glacier lakes still exist in this region. Much of this flat land, after the last Pleistocene glaciation of about 2 million years, which lasted till about 25,000 years ago, is now the breadbasket of Canada with major cultivation of wheat and mustard besides also supporting tree flora like spruce, balsam, willows etc. Other plants in this region include spear grass, blue grass, sagebrush, oat grass, larkspurs, lupins, hemp, asters, *Urtica, Clematis, Solidago,* etc. The altered natural habitat has many endangered animals with diminished populations of black bear, mule deer, elk, antelope, moose, squirrels, jack rabbits, beavers etc. This region also has many species of endemic grasshoppers, beetles and other insects, birds, fishes and reptiles. Despite having 80% farmland, only 10% of human population is engaged in farming and ranching. Most of human population lives in urban city centres and is engaged in services, mining etc. Several National Parks like Jasper National Park, Banff National Park, Waterton National Park, Yellow Stone National Park etc. were formed to conserve the flora and fauna along the Rocky Mountains in Canada and the United States. Despite such efforts, shrinking biodiversity due to freequent forest fires, loss of natural habitats due to urbanization, agriculture use and CO_2 pollution, mostly from burning fossil fuel result in green house gases which lead to global warming and unpredictable *El Nino* weather effects over the semi-arid parts of Asia. Friesen's (1987) book covers a wide range of topics on the historical and social perspectives of the Canadian Prairies.

Grasslands along mountain slopes in America, Europe and India are a fragile ecosystem prone to depletion, elimination due to soil erosion in rainy season and fire during summers. Report on Climate Change: Challenges to Sustainable Development in India (Occasional Paper (3). 2008), Rajya Sabha Secretariat, New Delhi states that mountain ecosystems are hot spots of biodiversity. The Himalayan ecosystem is considered as the lifeline not only to India but also to neighboring countries such as China, Pakistan and Nepal, owing to the perennial rivers that arise out of the melting glaciers. It is home to the largest collection of glaciers after the North and the South Poles. However, climate change is threatening this life giver drastically.

4.5.3 Flowering Plants and Insect Diversity

The beauty and grandeur of flowering plants together with that of bees and butterflies have enchanted humans ever since he took to evolve civilized societies. Flowering plants and associated insects, herbivore and other biodiversity having diversified largely since late Mesozoic, have met almost 90% of animal and human needs for food, fiber, fruits and vegetables, medicine and health supplements. They are constantly evolving in response to various selection pressures due to insects, herbivores, natural, agricultural and industrial toxicants. These are being constantly depleted due to human activity of unplanned urban habitats, deforestation, pollution etc. Not only the flowering plants but their partners in coevolution like insects, microbes and other biological resources are also disappearing rapidly. We are also losing large and diversified gene pool which has unlimited possibilities to enrich our food crops and fight many maladies afflicting human health like malnutrition, diseases associated with weak immune system like arthritis, cardiac and respiratory ailments. With such shrinking biodiversity of flowering plants and associated animal species, we have lost huge diversity and quantity of DNA which nature had meticulously evolved for the benefit of mankind. We are also losing a whole new diversity of yet undiscovered world of secondary plant substances and secondary animal substances which may have immense applications for sustainable agriculture, human health and nutrition. A Kew Garden report in 2013 has placed almost 20% of flowering plants as endangered species. When plant species get eliminated, most organisms along its food chain also become endangered. It can be safely concluded that saving plant biodiversity is the key to save total living biodiversity resources including the human resource.

4.5.4 Below Ground Biodiversity

Below ground biodiversity constitutes one of the largest reservoirs of biological diversity. These include species and billions of individual organisms being found within a single ecosystem, ranging from microscopic bacteria and fungi, through to larger organisms, such as earthworms, ants and moles. There are several kinds of beneficial organisms like nodulating bacteria that fix atmospheric nitrogen, antagonistic fungi like *Trichoderma, Gliocladium*, soil borne saprophytic fungi that keep pathogenic fungi in control, to earthworms that work through clay soils, add organic matter etc. Bardgett and van der Putten (2014) concluded in their review that below ground biodiversity of microorganisms and animals contribute significantly to shaping above ground biodiversity and the functioning of terrestrial ecosystems. They have pointed out that climate change affects soil biodiversity which affects soil fertility etc. Above ground and below ground components of communities and ecosystems have vital linkages (Wardle, 2002; Decaens, 2010). Relative to terrestrial biodiversity, the functioning of below ground biodiversity is less understood. Larger diversity of below ground biota will support complex food web and linkages with above ground biota that will eventually improve fertility and sustainability of soils.

4.5.5 Aquatic Biodiversity

World oceans cover 70% of Earth's surface, form the interconnected system of five major oceans and constitute the largest of sub-ecosystems on earth. Together with mangrove, estuarine flora and fauna along the coastal areas provide food, nutrition and livelihood to almost 40% of World Populations. This resource, despite being rich in biodiversity of marine invertebrate and vertebrate fauna, has faced massive destruction due to deliberate human activity of deforestation, mining, unsustainable harvesting of marine resources and pollution due to oil spills etc. There is general agreement that strong latitudinal marine diversity gradients, with declines in species richness to the poles, are more general for the northern than southern hemisphere (Clarke, 2009). Ancient climate change and evolution of antifreeze glycoproteins in notothenioid fish fauna have been linked to presence of several cryptic species (Near et al., 2012).

Chown et al. (2015) has shown that Antarctic biodiversity is more extensive, ecologically diverse and biogeographically structured than previously thought. It consisted of hexacorals, bryozoans, isopods, echinoderms, sea slugs etc. However, the impact of local human activities and global environmental change pose challenges to the current and future understanding of Antarctic biodiversity.

Besides, fresh water fish and other aquatic resources are a major source of food and nutrition for humans. Often runoff pesticides from agricultural fields, toxic industrial pollutants and even oil spills over marine habitats cause heavy mortality to such biodiversity resources. Aquatic ecosystems are more fragile than terrestrial ecosystems as they depend only upon green algae and phytoplanktons as producer organisms. If some of these were to mutate and develop bio-toxicities in response to selection pressures due to toxic exposures, life of entire aquatic biodiversity would be at stake.

Extreme climatic events, such as heat waves are predicted to increase in frequency and magnitude due to global warming but their ecological effects are poorly understood, particularly in marine ecosystems (Jentsch et al., 2007; Munday et al., 2008; Hegerl et al., 2011). In early 2011, marine ecosystems along the west coast of Australia - a global hotspot of biodiversity and endemism (Kerswell, 2006; Tittensor et al., 2010) experienced the highest magnitude warming event on record. Sea temperatures soared to unprecedented levels and warming anomalies of 2–4 C persisted for more than 10 weeks along more than 2,000 km of coastline. Wernberg et al. (2012) have documented that extreme climatic event alters marine ecosystem structure in global biodiversity hotspots. They have shown that biodiversity patterns of temperate seaweeds, sessile invertebrates and demersal fish were significantly different after the warming event, which led to a reduction in the abundance of habitat-forming seaweeds and a subsequent shift in community structure towards tropicalization of fish communities.

4.5.6 Human Natural Resource

Mother Nature created Man as its supreme natural resource. His presence in almost every living ecosystem on earth has brought in unsustainability in an otherwise sustainable living system. More recently, human greed has led to the abuse and destruction of natural resources, destruction of forests and other ecosystems, pollution of air, fresh water and marine resources. During the last century and particularly more so during the past few decades, more humans have been killed by humans due to avoidable wars, inter-sectorial rivalries, other man-made and natural calamities than almost the total present human population on earth. The rate at which human resource is getting depleted largely due to human activity is now far beyond the biological means of Mother Nature for their renewal.

This reminds me of satirical parody '*Brave New World*' published in 1932 by the eminent evolutionist, Alduous Huxley wherein he created a whole new world of humans with different levels of intelligence, like alpha plus, beta, gamma, epsilons etc. raised through the culture of test tube babies. When a reporter asked Mustafa, the chief controller, why he did not create everyone with alpha plus level of intelligence, he replied that they did create humans with alpha plus intelligence, put them all on one island and gave them all the resources. Very soon, they eliminated themselves, due to rivalry, infighting etc. Thus, it may be said that perhaps we may have become super intelligent biological species, that one day we may become the cause of our very own elimination.

With better acquired intelligence, it is expected that human civilizations will adopt equitable sharing of natural resources, irrespective of caste, creed, gender and social status. If we have the will, this world is not short of resources to meet this obligation. Respect of social system, respect of women at home and in society, equal opportunities in education and employment will automatically bring down the rate of population growth.

4.5.7 Biosphere

Life on Planet Earth evolved more than 3,000 million years ago. Humans have evolved in less than 50,000 years ago. The World human population has more than doubled since mid-1960s. The world human population which is currently 7 billion is likely to increase to 8 billion by 2025. The rate at which biological species of flowering plants, vertebrates and other animals are now disappearing is several times higher than the normal. Eastes et al. (2011) has shown how extinction of top carnivores' impacts ecosystem processes, which emphasizes the urgent need to integrate BEF and BES researches.

World oceans cover 70% of Earth's surface and constitute the largest ecosystems on earth. Together with mangrove and estuarine flora and fauna along the coastal areas, it provides food, nutrition and livelihood to almost 40% of world populations. Coastal marine biota inhabits eco-sensitive zone, some of which show salt and fresh water adaptations in estuarine environments, use vast stretches of coastal marine habitats for breeding purposes. This resource, despite being rich in biodiversity of marine invertebrate and vertebrate fauna, has faced massive destruction due to deliberate human activity of deforestation, mining, unsustainable harvesting of marine resources and pollution due to oil spills etc.

Man is now the dominant, if not the most numerous member of every living ecosystem which together constitutes the biosphere. Since 1960's, increase in food production due to HYV's of green revolution crops and increase in populations, he is living far beyond his biological means, the rate at which nature renews the natural resources he consumes. It is still possible to evolve life styles to live within the biological means. If we can explore, conserve and make sustainable use of natural biodiversity resources, improve living habitats, improve energy resources, evolve sustainable agriculture, we can still ensure equal access to basic food, health and social security needs for the entire human population.

Chapter 5

Biodiversity and Agriculture

5.1 BIODIVERSITY AND AGRICULTURE

It is difficult to imagine how our precious plant genetic resources could get conserved without the diversified insect pollinator's fauna. Plants on their part are investing major part of their energies in producing scents, nectar, honey, brilliant colours, and of course, excess foliage and biomass primarily to sustain this intimate specificity of their relationships with select insects (Fig. 14).

Fig. 14. *Bee Pollination in Mustard Flowers*

Ever since late Mesozoic or Cretaceous, the Angiosperm flora during their reciprocal evolution, because of multiple selection pressures due to feeding by phytophagous insects, herbivores and abiotic stresses, diversified into the rich genetic biodiversity of more than 240,000 known species of flowering plants which provided the most diversified base for agriculture.

Most crop plants produce far more foliage and plant biomass than the minimum necessary to sustain optimum levels of productivity. Insect feeding in a diversified natural ecosystem has never been the cause of unsustainability. Insects knew this more than 300 m years before us and thus coevolved a rich and sustainable genetic diversity of flowering plants and insect species which passed the most stringent tests of Darwin's struggle for the survival of the fittest.

For ages, insects made liberal use of the rich diversity of secondary plant chemicals as cue chemicals which guided them to their preferred food plants, which eventually became their host-plants. Phylogenetically related phytophagous insects evolved on phylogenetically related plant species and thus there were crucifer adapted, legume adapted species and so on. A few members of the Orthopteroid orders like the locusts and grasshoppers retained the primitive polyphagy, but most phytophagous insects evolved narrow host-plant specializations of monophagy or oligophagy as shown in butterflies and leaf mining flies (Ehrlich and Raven 1964; Sehgal 1971a, b; Spencer, 1990).

Restricted feeding in nature on botanically related plant species or on unrelated plant species having similar secondary substances is probably the result of numerous evolutionary adaptations of phytophagous insect to the allelochemics (Whittakar and Feeny, 1971), allomones and kairomones of the host-plant(s). In several of the highly evolved insect orders, many species like termites, aphids, jassids, white flies, army worms, boll worms, fruit flies, leaf mining flies, butterflies etc. have become secondarily polyphagous. Even polyphagous agromyzid pea leaf miner, *Chromatomyia horticola* (Goureau), show discriminatory preferences amongst pea varieties (Sen and Sehgal, 1993). Role of olfaction and neuro-sensory responses in insect-plant interactions in insects have been well studied (Sen, 1998).

Reciprocal coevolution of insects has been one of the prime causes of diversification of plants' genetic resources especially of flowering plants. They are still working hard, without protests or strikes, in bringing about pollination and maintaining the hybrid vigour of such diverse plant foods and flowering plants. Plants on their part are investing major part of their energies in producing scents, nectar, honey, most brilliant colours, and of course, excess foliage and biomass primarily to sustain this intimate specificity of their relationships with select insects. Butterflies and plants provide good example of this where plants elongated their corolla tubes to match the length of butterfly proboscis or vice-versa. Besides scavenging the organic waste, insects provide valuable products like honey, silk, lac etc., whose value in commercial terms would be far in excess of any human estimation of the losses caused by insects to agricultural produce.

5.1.1 Production Systems

As agriculture meets 90% of calorie intake and fisheries another 10%, sustainable strategies of agriculture, livestock production and fisheries are essential to food security. International Assessment for Agricultural Knowledge, Science & Technology for Development, IAASTD (2009) statistics suggest that 2.6 billion people are engaged in agriculture and fisheries for their living. Remote sensing surveys have revealed that >20% of cultivated land has decreasing productivity, due to climate change, soil erosion, competition between agricultural land and biofuel, water etc.

For ages, our diversified agriculture production systems had given sustainable production and productivity which met everyone's needs, because these were largely Multi Species Life Support Systems (MSLSS). Such systems consisted of large number of locally adapted genotypes of agro-horticultural crops, which had coevolved their very own natural defences or Multiple Stress Tolerance (MST) in response to multiple selection pressures due to various biotic and abiotic stresses. Such genotypes, when grown in diversified mixed and multiple cropping systems supported rich biodiversity of herbivores, primary, secondary and tertiary carnivores and of course man, in an interconnected web of food chain limited only by the energy input at the primary producer level of green plants. Ecosystems, with larger biodiversity and complex food webs are more stable. In agro-ecosystems, plant species diversity is replaced by larger number of individuals of the desired crop. Wood and Lenne (2001) have shown that natural ecosystems can offer attractive models for sustainable crop production.

Our natural ecosystems are self-regulating largely due to the bio-regenerative biocycling of natural bio-geological resources by Multi Species Biodiversity. The only major drawback of such Multi-species Life Support System is high biomass/productivity ratio and its very large size. A diversity index of plant and insect species in any natural ecosystem has direct relationship with the stability and thus are best indicators of sustainability. Larger numbers of insect species per unit area often mean fewer individuals per species, hence fewer problems of insect pest epidemics.

5.1.2 Sustainability *vs* Unsustainability

All biological populations, including those of insect pests, mites, nematodes, microbes and other animals are part of diverse natural resource. Their populations are regulated by innate processes like competition, environmental and incidental mortality factors etc. Any disruptive pest population represents our failures in natural resource management. Unsustainable agricultural production systems together with unsustainable management of biotic stresses have been responsible for several of the recent crop failures. Sustainability in agriculture is not an option now; it is a biological necessity (Sehgal, 1999).

A single intervention in standing crop with toxic pesticides and weedicides kill millions of natural biological control agents, besides killing equally large number of innocent visitors which came to rest in the agro-ecosystem. As insects and plants are evolving resource, they respond by developing resistance which necessitates more toxic applications. As various issues of environment, food security and poverty are interlinked there is need to harness synergy in a science-nature-farmer interaction to improve, food, nutrition and livelihood through sustainable development of agriculture (Singh, 2002).

Ciegis et al. (2009) have shown that the concept of sustainable development and its use for sustainable scenarios has multidimensional ramifications, which combine efficiency, equity and inter-generational equity on economic, social and environmental ground. They concluded that the most appropriate definition of *sustainable development* is provided in the report of the *Brundtland Commission*, stating that "*Sustainable development is the development that satisfies the needs of the current time period without jeopardizing the ability of future generations to satisfy their needs*". This view is in conformity with its use in ecological and sustainable agricultural development. This means the populations and ecosystem exploitation use should remain within the nature's means of biological renewal, without jeopardizing the interests of future generations.

Wiek et al. (2015) reviewed sustainability in action, based on three case studies on disaster recovery, bioenergy and precautionary purchasing.

They presented and compared three sustainability science studies on the reconstruction after the 2011 triple-disaster in Japan; limited energy and livelihood options in rural Africa; and toxic chemical dispersion in San Francisco. They have concluded that sustainability science is collaborative development of solution options by creating actionable knowledge. This deviates from traditional academic pursuits in being problem and project-based learning, addressing real-world problems that range from high-risk technology disasters to persistent poverty, violent conflicts and over consumption of public goods, sustainability challenges, the call for profound changes in natural resource management, conflict resolution, conduct of business, valuation of consumer goods and the process of technological innovations. The three case studies presented show various applications, successes and challenges for sustainability science. Future research areas include: (1) implementation and evaluation of sustainability competencies in sustainability programs; (2) development of long-term trust and relationships with relevant stakeholders; (3) coordination of solution-oriented research efforts among academic, governmental, business, and public actors within explicit political economies; and (4) construction of institutional funding and support appropriate to the specific needs of sustainability science. It must explore politics and power dynamics to negotiate solution options that address the questions - sustain what? for whom? for how long? At what cost?

Conservation and management of natural biodiversity in and around diversified cropping systems will promote Multi Species Life Support Systems, besides help minimise pest populations. The larger the diversity of insect and other bioresources, the lesser will be numbers per species population. Thus, there may be no pests. Promoting suitable agroforestry together with crop diversification with aromatic and medicinal crops will increase availability of natural biological control agents.

Our present unsustainability in agriculture is largely due to the difference between the potential yields and the average on farm yields in most crops. Saxena (1987, 1990) had highlighted the differences between on farm and potential yield in chickpeas. He attributed this gap to inappropriate production practices, biotic and abiotic stress management, low plant stand and depleting soil moisture. He suggested a small starter dose of nitrogen and soil moisture as critical to better yields. Blade et al. (2002) had supported management of agricultural biodiversity and crop diversification in the Northern Great Plains, besides initiatives to promote legumes, cassava and maize for sustainability in Africa.

How can we bridge this gap between potential and on farm yields? Multiple window dressing approach to cultural constraints, pathogens, insect pests, nematodes etc. as prompted by post green revolution era of post 1960's may not restore sustainability. A holistic approach on combinations of multiple stress tolerant genotypes with better plant architecture to utilize soil moisture grown in diversified cropping systems, minimum pesticide intervention in standing crop and management of natural biodiversity resources in and around the cropping systems is essential to sustainability (Sehgal, 2005). Role of insect pollinators in sustainable agriculture and yield has often been underestimated. Kevan et al. (1990) highlighted the pivotal role played by managed and native insect pollinators as a key constraint to the sustainability of contemporary agricultural practices. The economic value of such insects to pollination, seed set and fruit formation greatly outweighs that suggested by more conventional indices, such as the value of honey and wax produced by honeybees.

With expected increase in human population from the present 7 billion to almost 10 billion by 2050, our Agricultural Production Systems (APSs') will come under constant pressure to sustain productivity at least at twice the current levels. Hybrids under development in several crops may provide some reprieve from food shortages. Transgenics, an attempt to introduce one or the other toxin producing genes in crop cultivars to make them unpalatable to major insect pests is not the preferred strategy, it may give only short term reprieve. But such crops must prove their utility and acceptability on farmers' fields. As humans, we also would like to know the long term effects once such toxins are introduced in our food chains. Will insects not evolve a new set of mixed function oxidases to enable them to feed on such protected plants and in the process, become even more dangerous super pests? Will all this confer the much needed sustainability in our largely monocultural agriculture?

In search for all these answers, we often seem to overlook the real problem, *i.e.*, the lack of Multi Species Biodiversity to sustain the Multi Species Life Support System (MSLSS) as discussed earlier. Such APSs' can still become sustainable if our improved crop cultivars can be grown as part of the mixed/multiple cropping systems, which will ensure larger insect biodiversity to sustain the self-regulating food webs and trophic levels necessary for sustenance of the bio-regenerative bio-cycling of natural resources.

Since 1960's, enhanced food production due to adoption of HYV's of Green Revolution crops provided the much needed but only temporary reprieve from food shortages. This food security was so great that it triggered population increase in geometric progression of the kind predicted by Malthus (1798) in his famous essays on population. As human population crossed the threshold limits imposed by such self-regulating APSs', modern agriculture with monocultures of high yielding but largely susceptible crop genotypes grown over large areas under heavy subsidies of irrigation, chemical fertilizers and pesticide umbrella, began to replace the locally adapted genotypes and this ushered in the Green Revolution during the early sixties. During the next three decades, such high yielding genotypes completely replaced the locally adapted stress tolerant, slightly low yielding cultivars in almost all crops. Pesticide umbrella soon developed big holes, as insecticide resistant insect pest populations appeared in most crops. Such populations in the absence of competition from multi species biodiversity multiplied rapidly to epidemic levels and became the prime cause of unsustainability. Standing crop failures have been largely due to such insect pest epidemics in cereals, pulses, oilseeds, cotton, tobacco and several of the agro-horticultural crops wherever insecticide use has been high.

Cotton based cropping systems which use more than 50% of the total insecticide used in agriculture, has had maximum crop failures due to the resistant boll worm populations in the states of Andhra Pradesh, Telengana, Karnataka, Maharashtra, Haryana, Uttar Pradesh and Punjab. Such crop failures have forced several farmers in these states to resort to suicide deaths. Of late, plant breeding has placed greater focus on breeding genotypes with tolerance to one or the other biotic or abiotic stress. Such successes have been largely confined to fungal or bacterial pathogens and have given good dividends in terms of higher yields in wheat and other cereal crops. Despite this, several genotypes having resistance to one or the other major insect pest have been identified in most crops, yet breeding for resistance against insect pests has not made much headway. Optimism in development of Multiple Disease Resistance and Multiple Stress Tolerance in legume crops, by the turn of the century was expressed by Nene (1988). Development of crop genotypes with MST to major diseases and insect pests and also to abiotic stresses, still appear a distant goal.

Since the 1960's, with our pre-occupation with green revolution, centuries old disciplines of ecology, systematic botany and systematic entomology/zoology as related to agriculture, forestry and animal sciences in agricultural and forestry education establishments were deliberately phased out of relevance. As human need for habitation, urbanization, food and fibre grew, this became a direct threat to our precious natural biodiversity resources that ensures the very survival of life. Agricultural establishments conveniently forgot that this diversified biodiversity, evolved over millions of years, also requires continuous exploration, conservation and propagation to confer sustainability not only to agricultural, horticultural and livestock production systems, but to the entire ecosystem which supports health and all living biological systems. Although a biodiversity rich country like India has more than 30 National Bureau of Plant Genetic Resources devoted to almost 10 species of food crops, yet it does not have a single national institute for holistic study of the most precious natural resource of life, the living biodiversity resources which includes plants, insects, animals and socio-economic development of human populations. Such studies on biological systematics involving speciation in populations of plants, insects and animals and their exploration, inventories, conservation and propagation in time and space have assumed much larger significance. This is a highly specialized field which requires lifetime of dedicated work to show case the dynamic system of biodiversity.

Nature, when left undisturbed, has always been the greatest creator, propagator, conservator and curator of living natural resources. Despite the Earth Summit at Rio de Janerio where more than 150 countries signed a pledge to conserve chemical, physical environment and biodiversity, little has happened on ground in most of the developing countries to show that anthropomorphic degradation of natural living and physical earth resources has slowed. This is more so in Tropical and Semi-Tropical environments, where the diversity of living animal and plant species is maximum.

5.2 WORLD AGRICULTURE

World agriculture including livestock production is at cross roads. Agriculture is the primary source of food, but modern practice of intensive agriculture affects environment, contamination of drinking water, soil degradation, erosion and reducing biodiversity. Sustaining agricultural biodiversity is essential for a sustainable improvement in food and nutrition security (Frison et al., 2011).

We have spent over a generations' time since the early 1960's adoption of HYV's of green revolution. Productivity gains, though less than those in USA and Japan, were indeed impressive enough to provide food security. Several of these human afflictions were related to unsustainable levels of pesticide residues in food and dairy products. Losses of biodiversity due to shrinking of natural forests, damage to ecology, pesticide runoff etc. were irretrievable and a major factor that caused unsustainability of living ecosystems. It is high time we introspect and compute gains we made and losses we sustained in this process of sudden transition from traditional agriculture that sustained us over 10,000 years and chemical agriculture of last few decades since 1960's onwards. Several questions come to mind. Did our ancestors, dependent as they were on traditional agriculture and dairy farming, lead a less healthy life? Could we not achieve similar or slightly less productivity gains by adapting HYV's in our traditional agriculture using natural and farm fertilizers, without the toxic chemical inputs? Could we not increase productivity of MST local genotypes through hybridization alongside? Could MST traits not be transferred to HYV's for sustainability? Was there a critical shortage of agricultural land that we had to resort to massive clearing of natural forests? Could we not increase enough productivity by integrating HYV's into dry land crops in semi-arid environments of tropical and sub-tropical regions? Could we not harvest surplus monsoon rainwater, a key input available free for crops in semi-arid environments using massive watershed approach? Were we not aware of timely warning by Rachel Carson's 1962 *Silent Spring* and US Think Tank that led to a ban on DDT in USA? Why basic principles of ecology and environmental sciences were ignored while adopting - read not adapting, chemical agriculture of green revolution HYV crops? Several of these questions that arise in retrospect must be answered before we decide to carry on this legacy of chemical agriculture. An in-depth analysis reveals the following facts. The bigger question is to be as we are, or not to be as we are now and opt for a change. There are clearly two major phases in world agriculture. One, the centuries old traditional system of crop protection with the use of botanical materials such as neem, nicotine, pyrethrin, rotenone etc. together with the use of MST varieties evolved through centuries of natural selection. For ages, man depended on such eco-friendly approaches which also ensured the sustainability of their diversified cropping systems; second, chemical agriculture which started with the discovery and application of DDT and other chlorinated hydrocarbons during early 1950's. To begin with, we must delink our mind set of adaptation of HYV's from adoption of green revolution technologies of high inputs of irrigation, chemical fertilizers, weedicides, chemical pesticides etc. to sustain high yields. Nothing less than a total paradigm shift will be necessary for this change which may be opposed by vested interests which exploit farmer needs for big profits and by the corporate sector, which may need to redefine their role for more sustainable technologies.

A recent UNEP (2012) report on "Avoiding Future Famines: Strengthening the Ecological Foundation of Food Security through Sustainable Food Systems" compiled by an eminent group of experts suggests that sustainable food production and dietary habits that fight malnutrition, improves nutrition and health besides also supporting green economy by sustainable use of natural resources at levels that can be renewed by biological means. It is estimated that 1.3 billion tons of food produced for humans is lost or wasted annually. In developed countries, 40% food is wasted at consumer or retail, while in developing countries, 40% is lost at post-harvest storage and processing levels. Despite impressive increase in agricultural production to meet demand of growing populations, malnutrition affects large portions of world population. The UNEP 2012 Report also highlights the significance of sustainable ecological foundations of agriculture. They have defined ecological foundation as the natural stocks and the flows that support the production of food. This is made up of two components; first, the resource base of agriculture like water, land etc. and the second, consisting of human consumption, cropping systems, livestock farming, pesticides, irrigation, mechanization, synthetic fertilizers, plant nutrients, nutrient cycling, climate, off and on-farm biodiversity that support the ecosystem. Amongst these, they stated that variety and variability of animals, plants and micro-organisms at the genetic, species and ecosystem levels are necessary to sustain key functions of the ecosystem. They have also endorsed the general perception that the more diversified the agricultural land, the more resilient it is to climatic and other disturbances. Diversity on the farm also helps maintain the genetic pools of plants and animals. Studies by Altieri (1999) and Kremen and Miles (2012) also show that diversified farming systems are capable of supporting substantially greater biodiversity, soil quality, water and energy use efficiency, carbon sequestration and resistance and resilience to climate change than simplified farming systems.

The World Food Summit (UNEP, 2012) defines food security as "*all people, at all times, have physical, social and economic access to sufficient, safe and nutritious food to meet their dietary needs and food preferences for an active and healthy life*". However, this did not happen because current world food production system ignored its ecological foundations due to neglect of natural resources of soil, water and biodiversity and also due to environmental pollution.

World agriculture is hardly 10,000 years old when tribal and nomad human populations turned to the wide diversity of angiosperm flora. Soon they discovered that the seeds they collected could germinate into new plants that would bear fresh fruits and seeds on maturity. Biological necessity for survival forced early man to look to this vast natural resource of flowering plants for food. The need for cultivation, propagation, conservation and improvement of selected plant species considered beneficial forced him to stay at one place to harvest the produce, which led to the development of communities, cities and even countries. This necessitated settlements along major waterways of the world to sow seeds and nurture crop plants to maturity. That was the beginning of most world civilizations like the Amazon, Greek, Nile valley, Indus valley, Indo-Gangetic plains, Brahmaputra river valley etc. Corn was perhaps amongst the first crops to be grown in such systems and was followed by several other major crops. Such systems consisted of large number of locally adapted genotypes of agro-horticultural crops, which had evolved their own natural defences or Multiple Stress Tolerance (MST) in response to multiple selection pressures due to various biotic and abiotic stresses. In agroecosystems, the plant species diversity is replaced by larger number of individuals of the desired crop. Improved agronomic practices could easily overcome any such constraints in productivity with precision application of required natural resources.

5.2.1 World Trade Organisation (WTO): (Policy Perspectives & World Agriculture Trade)

In the recent past, export/import of several crops including mangoes, vegetables like brinjal, tomatoes etc. to and from UAE, Japan, EU, and even USA has declined due to presence of high levels of pesticide residues and insect infestations.

In the foreseeable future, world trade in agriculture, agrochemicals, pesticides, pharmaceuticals, health products and services will remain essential to sustain demands for world population of humans and livestocks. Siddiqui (2015) citing WTO reported that in 2013, global agricultural trade totalled US $ 1.7 trillion. Agricultural trade and tariff practices are often subjected to controversies due to subsidies, protectionist policies, sanitary, phytosanitary issues etc.

World Trade Organisation in 1995 had set minimum standards of protection and Intellectual Property Rights (IPR). The three main features of the TRIPs (Trade Related aspects of Intellectual Property Rights) agreement cover issues related to Standards, Enforcement and Dispute settlement.

- ***Standards*:** A patent is given for a term of 20 years. Plants, animals and other organisms may be excluded from this regime. Plant varieties are to be protected by the patent.
- ***Protection:*** Provided for undisclosed information, trade secrets etc.; prohibits disclosure of test data regarding new chemicals and products and protection of test data.
- ***Enforcement:*** Permits effective action against violation of Intellectual Property Rights (IPR), avoids creation of barriers to legitimate trade; criminal proceedings may be initiated for infringement of IPR.

To be a part of world trade, every country has to go along with WTO to avail opportunities under the TRIPs agreement for developing, patenting, setting quality standards for new technologies. Similar views to support WTO and IPR rights were expressed recently by Ambassador I.A.Siddiqui (2015), Senior Advisor, Global Food Security Centre for Strategic & International Studies (CSIS), Washington, DC, USA at Pantnagar, India. For latest updates refer UN's WTO website.

5.2.2 Legislative Support: (Environment Protection Agency (USEPA) in USA, UNEP and the Indian Biodiversity Law 2002)

EPA of United States is an agency of the Federal Government created for the purpose of protecting human health and the environment. This Agency was proposed by President Richard Nixon and became operative on 02 Dec. 1970 with headquarters in Washington DC and regional offices in 10 regions and comprises 27 laboratories. The EPA sets standards and has enforcement powers.

USEPA has bilateral cooperative program with several countries that include Australia, and the countries in the Asia Pacific region which include China, India, Indonesia and Japan, Singapore, Thailand and Vietnam. Besides protecting human health, the agency also cooperates in controlling air pollution and help in clean energy economies. United Nations Environment Protection (UNEP) has active programs for protecting human health and reducing pollution in the world.

The Indian Biodiversity Act 2002 requires all states, cities, municipalities and village panchayats constitute Biodiversity Management Committees (BMC) and maintain People's Biodiversity Registers (PBR). Hindustan Times India Report August 12, 2016 highlights that most states except Kerala, Karnataka and Odisha have done little to constitute BMC's and prepare PBR's. PBR's compile record of traditional knowledge, medicinal value or any other use of all biological resources and make people aware against commercial exploitation of such resources. India has only 2.4% of land area and 4% of water. It accounts for almost 8% of biological species recorded in the world that includes 45,968 species of plants, 91,364 species of animals and 5,650 species of microbes (Source: RTI responses from states and National Biodiversity Authority). Killing tigers for trading its bones and clearing *Chilgoza* pines in forest lands for industrial projects are a few examples where people lose out. This requires documentation of authentic taxonomic records of all biological species and their value to mankind to create awareness amongst people. Despite available safeguards and checks, biodiversity losses continue unabated in all world biodiversity hot spots and eco-sensitive zones.

5.3 GREEN REVOLUTION: BREAKING PRODUCTION BARRIERS

Eminent plant breeders from USA helped identify wheat and rice varieties from USA and Japan that were being grown under high input technologies of chemical fertilizers, irrigation, weedicides and pesticides etc. The credit for developing HYV's of wheat and rice and thereby the beginning of the Green Revolution goes to Dr. Norman Borlaug, the Eminent American Scientist and Plant Breeder, who, while working in Mexico had pioneered the adoption of such varieties not only in USA, but also in far off thickly populated regions of India and other Asian countries. These varieties had the potential to double grain yields in south-east Asia and elsewhere. In India, such varieties were adopted from early 1960's onwards. Since 1960's onwards, realising the importance of Agriculture in India, the then Prime Minister Jawaharlal Nehru facilitated the establishment of a chain of State Agricultural Universities, largely through India-US collaborations. This accorded much needed prestige to the profession of agricultural science and to scientists engaged in scientific pursuit of agriculture. True to its potential, adoption of Green the Revolution HYV's transformed world agriculture since early 1960's wherever such varieties were adopted. Wheat and rice yields almost doubled, this brought relief to large populations in the developing world. The successful transition from 'ship to mouth' to 'field to mouth' was hailed by Dr.M.S.Swaminathan, an eminent plant breeder and geneticist and his team in India and was also applauded the world over for bringing freedom from hunger.

Today, world agriculture produces enough food to feed world population of almost 7 billion. At present, agriculture in thickly populated countries as India depends heavily on applications of costly inputs of chemical fertilizers, weedicides and pesticides. In an environment largely unmonitored for pesticide residues, people consume unacceptable levels of such toxic chemicals on their dining table. Deteriorating national health, cancer and diseases due to weak immune systems are enough evidence of this.

5.4 BIOLOGICAL WARFARE BEGINS: ALL LOVE IS LOST

Linking HYV's with high inputs of chemical fertilizers and irrigation and the pesticide umbrella concept of USA and Japan involving scheduled applications of pesticide chemicals was an avoidable ecological error. Green Revolution, as it is associated with chemical agriculture included high input technologies for application of irrigation, weedicides and a scheduled application of pesticides to achieve the desired yields. Very soon, the romance was over. What was supposed to be the saviour from human hunger turned out to be a biggest ecological disaster, more so in thickly populated countries such as India, China, Africa and Latin Americas. Resurgent insect populations in such fields could have been managed with other options of effective seed treatment technologies, botanicals, fumigants, other ecological options of agronomic manipulations and incorporation of resistant traits from locally adapted MST varieties.

Within five years, the number of pesticide companies proliferated from single digit to over five hundred. The discovery of chlorinated hydrocarbons like DDT, BHC in early forties had already changed the entire scenario and their application in agriculture soon became the universal cure for all insect and mite problems in crops, storages, homes, on livestock and even for body louse control on humans. As insects developed resistance to DDT, more deadly options of organochlorines, organophosphates, carbamates and synthetic pyrethroids were introduced. Because of poor regulatory control, aggressive marketing for quick profits by the pesticide industry, often through dealer/money lender exploitation of farmer's needs, a very intense selection pressure was placed on insect populations.

Selection pressures due to such large scale use of highly toxic pesticides were so great that it led to wide spread killing of diversified insect species fauna and several of the natural biological control agents which used the crop field as their resting habitat. This not only included plant feeders, but also the parasitoids and predators and other species including birds which provided the inbuilt natural regulatory mechanisms of species population control. A few such species populations which could overcome the selection pressures multiplied their populations, free from competition, rose to levels where they again became pests. This not only led to the resurgence of new pest populations, but also to the ultimate development of wide spread insecticide resistance in insect populations. This is what opened the Pandora Box. Agriculture since then was never the same as the system evolved before by nature during several hundred centuries.

5.4.1 Agroecosystems and Pesticides

Agroecosystem, like any other ecosystem are life's habitats. As such, pesticides and ecosystems are not biologically and ecologically compatible to life and living resources. All chemical pesticides are biological poisons. There is no such thing as safe or unsafe chemical pesticide. Agroecosystems are used by a large diversity of local flora, organisms which include microorganisms, several invertebrates and insects, parasitoids and biological control agents, birds and mammals. Most of this natural biodiversity resource consists of innocent visitors, who happen to use the crop habitat as a resting place. Others are harmless feeders on crop and weed flora. Several insects and spiders as biological control agents feed on other insects that fed on crops and weeds. This complexity of natural biodiversity sustains a complex food web of producers, herbivores and carnivores which provide natural control of biological populations as multi-species life support system. Strategies for crop management should primarily focus on crop plant health to sustain optimum productive yield, rather than on killing all biological populations with chemical means. Chemical pesticide intervention in any standing crop should be minimised as it would kill the vast diversity of beneficial resource as well as innocent visitors to crop, thereby affecting sustainability adversely. Biological populations should as far as possible be managed by biological means. Host-plant resistance and conservation of natural biological control agents, together with cultural management of crop health should be the preferred options. Chemical means only delay the solution. Biodiversity of weed flora often serve as trap crop for several insect populations that would otherwise feed on crop plants. Weeds like *Chenopodium* and *Vicia* in chickpea fields in cooler north India environments attract larger oviposition by *Helicoverpa* moths than chickpea leaves which have acid exudates. Emerging larvae migrate to chickpea pods (Fig. 15). Removal of such weeds just prior to pod formation in chickpea fields would control pod borer damage.

Why pesticide residue damage is severe in thickly populated countries as India and other Asian countries than in relatively thinly populated countries as USA and Canada and in not so thinly populated China? After the Second World War and later due to struggle for independence, most Asian countries inherited a damaged ecology. Thickly populated countries as India had low forests to agricultural land ratios compared to that in thinly populated countries and even in China. This large biodiversity of natural forests absorbed more harmful gases and chemicals, also released more oxygen to the environment which reduced much of the harmful effects of pesticide residues on humans. Besides, developed countries have better means of monitoring and environment protection.

5.5 PESTS: INSECTS, WEEDS AND MICROBES

Nature does not classify biological populations as pests, weeds or microbial pests etc. All these are anthropomorphic labels given by man. No biological population, which has evolved through millions of years of selection pressures and Darwinian struggle for the survival of the fittest can be trashed as pest or weed. Since 1960's green revolution, crops responsive to mechanised technology, chemical fertilizers, weedicides and pesticides, biological populations came under severe selection pressure of toxic chemicals. Though food production and populations increased rapidly, this brought along with it several ecological disorders. Soon we had a long list of insect pests, weed pests, animal and nematode pests, microbial pests, fungal pests etc., which further triggered large scale proliferation of killer chemical pesticide industries. Despite belated realization of damage being caused by applications of toxic chemical pesticides, the concept of Integrated Pest Management remained largely unimplemented even till today. Aggressive marketing and government subsidies have ensured consistent annual increase in the use of chemical pesticides and even supported unsustainable aerial application of chemical pesticides. Pesticide industries and agricultural establishments misled farmers about safe and unsafe pesticide applications.

Fig. 15. *Chickpea Pod Borer, Helicoverpa armigera larva feeding on Chickpea Pod*

Biological populations are not born pests, rather, like humans have evolved over millions of years through natural selection to occupy a specialized ecological niche. Chemical pesticides that kill biological pests are equally toxic to humans, livestocks and biological control agents. To prevent this misuse, World Agriculture and FAO set maximum pesticide tolerance limits for human and livestock foods. Lack of monitoring and certification at district and village level has meant that people mostly in developing world continue to consume high levels of pesticide residues.

Most insects and pests which fed on treated crops developed varying degrees of resistance, necessitating the use of more toxic options for pest control. Pesticide residues are now present in most foods we consume including fruits, vegetables, dairy products, even in mothers' milk. The Bhopal gas tragedy on 2-3 December, 1984 in the early hours of morning where thousands of people died due to pesticide gas leak from Union Carbide plant at Bhopal, India is considered as the world's worst industrial disaster. This also provided enough evidence of lethal toxicities of chemical pesticide use on humans. According to Wikipedia, over 500,000 people were exposed to methyl isocyanate (MIC) gas and other chemicals. Real damage is still happening in urban and rural populations, where thousand fold increases in lung cancer, respiratory diseases, cardiac problems etc. have claimed millions of lives largely due to food laced with pesticide residues and toxic environmental pollution. Large populations in developing countries do not have any kind of health security or access to quality medical care. Almost 60% of population consisting of small farmers and farm workers in these countries do not have any social security. Given the present minimum wage structure, quality food, horticulture and dairy are far beyond the reach of a majority of common people. This is despite the fact that almost 40% of food produced by hard work of small farmers is lost between harvest and mouth, due to lack of proper post-harvest handling, inadequate processing, storages etc.

5.6 INSECTICIDE RESISTANCE IN INSECT PESTS

Several of the long persistent insecticides or their toxic metabolites, entered the food chains in diverse ecosystems. Soon, toxic residues, often far above the permissible WHO limits were detected in agro-horticultural produce and even in such diverse commodities as milk, butter and in mothers' milk. The warning in Rachel Carson's' (1962) famous book '*Silent Spring*' forced many scientists to re-evaluate the role of chemical insecticides in sustainable agriculture. Besides several afflictions due to contaminated food and environment, we are now faced with the problem of insecticide resistance management in insect pests. Over 500 species of insects have become resistant to available insecticide options. Fortunately, our experience of working with resistant pest populations have shown that they soon return to susceptible status, when severe selection pressures due to chemical pesticides are removed (Mehrotra and Phokela, 1992). Despite the knowledge that pest populations do not occur at uniformly high level of biotic stress at all latitudes and altitudes, pesticides continue to be applied as per industry sponsored crop packages.

5.7 MAN MADE ECOLOGICAL DISASTERS

5.7.1 Clearing Natural Forests for Farm and Urban Needs

All in the name of Green Revolution, several thousands of hectares of natural forests in eco-sensitive zone of 'Terai' foothills along the Himalayas; major grasslands like Prairies in North America; in the catchment areas of mangroves in the Bay of Bengal, along Western Ghats in India, coastal areas in all continents and tropical rain forests world over were cleared, causing irretrievable damage to the world ecology resulting in a major ecological disaster.

These natural forests, evolved through natural selection over the last 60 million years of the rise of Himalayas due to striking, sub-ducting of the Indian plate with the Eurasian plate, not only had the rich diversity of tree flora of broad leaf Angiosperms, but were also the world famous heritage site of rich biodiversity including the famous Kumaon tigers and leopards. These natural forests also acted as a buffer zone protecting the sensitive ecology often disturbed due to high seismic activity of rising Himalayas and flash flooding of rapidly flowing rivers during monsoon season. Together with the rich diversity of dominant pine forests along the mountain slopes, these forests prevented erosion and the frequently occurring landslides. This further prevented silting of the rivers in the plains and rivers flooding villages and cities along the plains. Large tracts of these natural forests could have been easily spared as National Biodiversity Parks with limited human access for adventure and tourism. Perhaps, it would have earned much larger revenue than the gain in agricultural crops and urban needs. So great was this loss that future generations will never understand the magnitude, splendour and magnificence of these natural resources of plant and animal species diversity that these forests nurtured. There was enough land to meet human needs world over. Parts of such forest lands could have been used for agro-horticultural and dairying research to provide farm support to local populations, besides undertaking fundamental studies/research of national and world importance. Setting up of the first Agricultural University as U.P. Agricultural University, now known as G.B. Pant University of Agriculture and Technology at Pantngar, Nainital, India in the Kumaon region in 1960 was a welcome step in the right direction. The core structure of the '*Terai*' Forests in Himalayas could still have been conserved as national parks and for wild life studies.

In the absence of such forests now, man-animal conflicts, soil erosion in upper mountain regions, silting of rivers and frequent changes in their course, flash flooding of thousands of villages year after year during rainy season could have been partly spared. This has already claimed millions of human and livestock lives besides destruction of their homes. If left undisturbed, it will take at least couple of centuries to partially restore such forests.

The legendary British hunter turned conservationist and naturalist, Colonel Edward James "Jim" Corbett (1875-1955) had established his home at Kaladungi in the Terai Forests of the Kumaon district of Nainital in the central Kumaon Himalayas (Corbett, 1954). He was famous for hunting large number of man-eating tigers and leopards in India and author of a famous book '*Man-Eaters of Kumaon*'. In later years, he turned a conservationist and played a key role in creating a national reserve for the endangered Bengal Tiger. In 1957, a national park in Nainital district of Kumaon Himalayas was renamed as Jim Corbett National Park in his honour. Vast plains of northern India had for centuries provided sufficient resource of agricultural land and evolved rich diversity of MST genotypes of all major crops. Why Green Revolution crops could not be raised in such areas?

5.7.2 Valley Agriculture, Terrace Cultivation

Valley agriculture sounds music to ears and a perfect setting for Bollywood films with the backdrop of the majestic Himalayas. For residents, unplanned ecologically unsustainable development of this region was no less than an ecological disaster. Scattered along mid-hills in Himalayas are a few apple, pear and apricot orchards. Prominent among these and mostly established during the British India rule were Ramgarh valley, Ranikhet ranges in Kumaon, Chakrata orchards in Garhwal region, Kotgarh in Shimla and Himachal ranges and Kashmir orchards. These orchards now are mostly senile or house obsolete varieties, require large scale rejuvenation by grafting from root stocks of improved varieties now in demand the world over.

Unplanned cutting of pine forests all along southern mountain slopes for timber, resin and wood for making packing boxes and other human needs of housing etc. led to shrinking of natural forests to a few protected spots. Evergreen Himalayas largely turned brown due to grasses and other xerophytic vegetation encroaching on the vacated lands. Similarly, cutting trees along river valleys loosened soil, causing silting of rivers in hills and down in plains. Unsustainable terrace cultivation with wheat, rice, together with urban and commercial settlements along the banks of major rivers, brought disasters that led to frequent landslides, unseasonal rains, cloud bursts etc. Kedarnath, the high altitude site of Lord Shiva in the Garhwal Himalayas, unexpectedly received heavy rains due to a sudden cloud burst from June 14-17, 2013 causing unprecedented death and destruction in the region. Thousands of people died and several went missing besides destruction of property all along the route. Despite the magnitude of this disaster due to the sudden cloud burst, it highlighted how the ecology of the region was systematically eroded to serve human commercial interests. Houses, markets and resorts were built over loose soils on the banks of river after cutting and clearing of forest trees. All these, together with large human settlements were washed away within a few minutes of the onset of this disaster. Had at least 500 meters of river embankments been conserved with original trees and plantations, this calamity could perhaps have caused lesser damage.

5.7.3 Conservation of Himalayas and Similar Mountains as Natural Resource Parks

Palearctic species from the west and Mongolian species from the east have enriched temperate and even alpine regions in Himalayas. Over the years, many such species have evolved into endemic sister species, while others have extended their ranges to become Holarctic in distribution. Besides all these, many of the Ethiopian and Australian species have extended their range northwards to make foothills as their new home. Major explorations of Indian fauna occurred during the British rule, which are documented in a series of volumes of '*Fauna of British India*'. Sir J. D. Hooker's '*Flora of British India*', 7 vols. (1872-1897) published by Drury (Heber) is still a classic work on the diversity of flora in British India. Likewise, the '*Flora of the Upper Gangetic Plains*' by J.F. Duthie (1903- 1922) is another valuable reference.

The ecologically sensitive Himalayas, consistently prone to moderate to heavy seismic activity upto 7.0 and above magnitude, do not lend itself very well to agriculture of plains. The frequent earthquakes in the region perhaps do not kill as many people as the poorly developed infrastructure, housing and other structures. The most recent Nepal earthquake on 25 April, 2015 of 7-9 magnitude killed several thousands of people and rendered millions homeless. Most of the deaths were due to collapse of infrastructure and housing. It must be understood that human habitations and infrastructure development should be conducive to conservation of rich biodiversity and must be ecologically sustainable. The natural beauty and splendour of the Indian Himalayas, together with its valleys of alpine flowers and several pilgrimage destinations should be systematically developed to promote safe tourism as a major industry. These regions should have been developed for production and processing of value added horticultural crops and for propagation of value based medicinal, aromatic and flowering plants suited to the temperate and alpine climates. The Himalayas, like mountain ranges in Switzerland should also be developed for profitable production and processing of dairy and animal products.

5.7.4 Conservation of Mangroves

Mangroves forests and associated flora constitute an equally important biodiversity zone worldwide. This protects the rich diversity of marine coastal fauna and flora and also provides for habitation of salt and water tolerant flora and fauna of the ecologically sensitive habitat. The livelihood of millions of people along the coastal areas depends on the rich diversity of marine fauna and flora. Lately, ecologically unsustainable mining and cutting of mangroves for human habitations and commercial use have threatened the ecology and survival of this natural resource.

5.8 IPM-INTEGRATED PEST MANAGEMENT

As soon as we think of IPM, we put ourselves in a battlefield of defendant *vs.* enemy biological pest populations. Where do we start? What should have been a strategy for co-habitation of ecosystem became a battleground for biological warfare. No holds are barred. Chemical weapons of mass destruction often become the best and the first window dressing option. Though there is considerable awareness about harmful effects of chemical pesticides, their use and abuse continues unabated. Quick successes, easy availability, ready to use formulations have made chemical use a preferred option. This has also hindered progressive search and refinement of other options of host-plant resistance, botanical pesticides and biological control. Agricultural science should address this as a holistic approach for abiotic and biotic stress management. Thus, treating IPM as Integrated Biotic Stress Management (IBSM) seems more inclusive. Despite these above concerns, IPM is discussed as an ever evolving strategy for ecological approach to manage disruptive pest populations (Sehgal, 2001).

5.8.1 IPM-Concepts and Success Stories

IPM is typically an ecological approach for protection of plant health from all kinds of pest populations like harmful insects, animal pests, disease pathogens, weeds etc. It is essentially a farmer activity of using one or more management options to reduce pest populations below the economic injury level, while ensuring productivity and profitability of the entire farming system with emphasis on environmental sustainability and human resource development. Thus, in IPM, the main focus is on plant health and farmer interests, rather than the individual pest or pathogen.

IPM - Major objectives

- To ensure plant health and safety of our environment.
- To reduce our total dependence on toxic chemical pesticides to a minimum necessary to sustain optimum productivity.
- To identify strategies of pest control based on combination of farmer and cultural practices, including host-plant resistance, conservation and augmentation of biological control agents.
- To promote the use of biodegradable botanical pesticides like neem, karanj and other plant based products.

Why IPM?

The necessity to promote IPM must be viewed in the context of the urgency to produce sufficient quantities of healthy food from our agro-horticultural crops necessary to sustain our growing populations and to provide healthy environment for human resource development.

A quantum jump in the production of food grains achieved during the green revolution has now been completely offset by our population growth which is expected to cross 10 billion by 2050 AD. This has already reduced availability of per capita food grains to levels far below necessary to maintain the health of our people. Even at a modest estimate of crop losses due to insect pests and diseases at 30%, the estimated cost of food crops lost is more than approx. US $ 50,000 million at current values though India and several of developing countries are using much less pesticides than that in developed countries like USA and Japan, where pesticide use is very high. For example, 54% of the total pesticides used in agricultural sector is consumed in cotton based cropping system though cotton is grown only over 5% of the cropped area. Four states, Andhra Pradesh, Telengana, Gujarat and Karnataka consume 65% of the total pesticides used in India. More than 500 species of insect pests have already become resistant to one or more of the commonly used pesticides, leading to crop failure and even suicide deaths of the farmers in several Indian states like Gujarat, Karnataka, Telengana, Punjab, Haryana, etc. Milk and butter samples in several of the above mentioned States contain up to 2.4 ppm of DDT residues. Even mother's milk contain DDT residues at levels >2 times the permissible limit by World Health Organization. In the absence of any monitoring, fruits and vegetables consumed often contain pesticide residues far above the permissible limits.

Several key insect pests, because of their regular occurrence at levels above the economic injury level, cause economic losses worth several hundred million dollars in several crops. Major insect pests that farmers have to deal with is presented in Table 3.

Table 3. *Major insect pests of crops*

Type	Crops
Flower, pod, fruit, seed feeders	Pulses, cotton, sunflower,tomatoes, linseed etc.
Plant hoppers	Rice, sunflower, mango etc.
Aphids	Mustard, legume crops
Internal stem feeders	Cereals, pulses, oilseeds, sugarcane and several horticultural crops
Polyhagous pests like hairy caterpillars, termites, white grubs	Cereals, pulses, soybeans oilseeds, sugarbeet and hill horticultural crops

The present levels of productivity of several of the green revolution crops is being maintained because of farmers' adoption of HYV's, which are grown under high inputs of chemical fertilizers and pesticides. Largescale use, rather misuse of long persistence toxic chemical pesticides has made our agro-ecosystems highly unsustainable. This has also brought about degradation of our natural resources on land, air and water and the resulting environmental pollution caused has already affected human health adversely. Besides, this has eliminated several of the non-target organisms like pollinators and natural control agents from our agroecosystems. Consequently, we have faced crop failures due to resurgence of pest populations, appearance of several new and secondary pests and the development of insecticide resistance in natural pest populations. For example, the pod borer, *Helicoverpa armigera*, also known as American boll worm or gram pod borer has become a pest of international importance, causing economic losses to several crops like chickpea, pigeonpea, cotton, sunflower, tomato, chillies, tobacco etc. *Heliothis/Helicoverpa* species cause an estimated loss of US $5 billion annually; despite the fact that pesticides costing over US$1 billion are used to control them on different crops (Sharma et al., 2005). The states of Andhra Pradesh, Telengana and Punjab, where pesticide use is among the highest in India, has already had several epidemics of boll worms and white flies in cotton and pulse crops. High selection pressure, due to toxic chemical pesticides, led to widespread insecticide resistance in bollworm populations in south-east Asia and Australia, Africa etc. It was only after this wide spread failure of chemical control measures during the late eighties, the search for alternatives to chemical pesticides for the control of pest populations got some momentum. It is in this context, the relevance of IPM in agriculture has come to be appreciated not only by farmers and scientists, but also by agricultural policy makers.

Pest problem is always a population problem. In fact, a Latin world *Populari* means to ravage or to destroy. Populations are living biological entities and they present biological problems. They can be reduced by chemical insecticides but rarely eliminated. Thus, biological pest problems cannot be solved with chemical insecticides, though the solution of these problems may get temporarily delayed.

We must remember that several of the million species of insects also contribute to human welfare. The pollination in most of the cross pollinated agricultural and horticultural crops, honey, silk, lac etc. are only some of such benefits that we derive from insects. In fact, we are dependent upon other organisms, both plants and animals including insects, far more than they depend on us. It is also important to note that in ecosystems that have greater diversity of plants and animal species, there are far less fluctuations in populations of pest species, ie., there are less pest problems. One hectare of paddy (rice) field may have upto 10 million parasitoid and predators that will keep most pests in control.

Biotic Stress Mapping

Crop losses due to insect pests are known to vary considerably between years and geographic areas because of differences in agronomic practices and environmental factors (Matthews, 1984). Although information on Damage/Yield Relationships (DYR) is essential to establish Economic Injury Levels, this aspect has not received much attention by crop entomologists. Based on several years of accumulated data on the survey of major insect pests of chickpea and pigeonpea crops, the pod borer, *Helicoverpa armigera* and pod fly, *Melanagromyza obtusa* (Lateef and Reed, 1984; Sitanantham et al., 1984; All India Coordinated Pulse Improvement Project Annual Reports - AICPIP 1982- 1988) along with time dimensions added to the data from research stations and in on-farm situations, the relevance of estimation of yield loss and biotic stress mapping was emphasized (Sehgal, 1990a). The results confirmed the null hypothesis that biotic stress and yield losses due to *Helicoverpa armigera* and *Melanagromyza obtusa* are not uniform and show consistent pattern of variations across the latitudes in India. Such data when superimposed on distribution area and production pattern of chickpea and pigeonpea crops in India showed that large areas where biotic stress and yield losses were less than 10% did not require toxic chemical control.

In chickpea crop, *Helicoverpa* damage was high (>20%) in North-West Plain zone and in parts of Central zone, where there was perfect synchrony of peak pest populations with maturity stages of the crops. Pod borer damage was low (< 10%) in North-East Plain zone and in West zone. Differences in severity of damage between research station data and on farm survey data were probably due to the fact that on-farm samples were taken when the crop was close to maturity.

The pigeonpea crop is damaged by two major pod borer pests, pod borer, *Helicoverpa armigera* and the pod fly, *Melanagromyza obtusa.* While pod borer damage was generally high (>20%) in almost all areas, yet it was low in parts of Central Zone, CZ around Gwalior.

The relationship between pod damage and yield loss is linear, more so when damage occurs close to maturity at Pantnagar and Hisar in northern India (Pimbert et al., 1989; Pimbert and Sehgal, 1989; Sehgal, 1990 a, b; Sehgal et al., 1990). Reed et al. (1987) suggested that percent pod damage cannot be directly related to yield loss as foliar damage may reduce the number of pods that are carried by the plants. In major chickpea producing areas of northern India, little foliar damage occurs during vegetative stage due to low mean winter temperatures as at such 'hot spots' like Pantnagar and Hisar, the crop suffers substantial damage close to maturity.

The pod fly, *Melanagromyza obtusa* damage was severe in CZ and in SZ, while it was relatively moderate in NWPZ and in NEPZ. Such differences will require further analysis in relation to crop phenology and different maturity groups. The pod fly has two related alternate hosts, *Moghania macrophylla* (Willd.) O. Ktze which is cultivated around Ranchi in CZ as a host for the lac insect, *Laccifer lacca* (Kerr.) (Sehgal,1965) and *Flemingia congesta* (Wild) which grows wild in Kumaon Himalayas. These hosts aid in offseason carry over of pod fly populations (Sithanantham and Sehgal, 1985).

Yield loss estimations due to insect pests are essential across large areas of crop distribution. Out of large number of insect species associated with greengram crop in India, only few insect populations reach pest status in different regions and growing seasons (Sehgal and Ram Ujagir, 1987). Yield losses should be expressed as percentage of reduction in potential yield that would be obtained in the absence of pests. When pest damage is high upto 45% in cereals like rice, maize etc., there is little time for the crop to compensate; we often get a linear relationship with percentage yield loss. When pest damage is low or occurs early in crop phenology, plants tend to compensate the damage in different degrees.

IPM - How does it work?

IPM is based on critical monitoring of the pest populations to identify intervention points that will prevent the general equilibrium of pest populations from reaching the economic injury levels. Development of pest forecasting and forewarning systems based on simulation models of pest populations also help farmers adopt timely control measures based on combinations of farmer and cultural practices including host-plant resistance, use of natural and exotic biological control agents and minimum need based insecticide use. All this is not so easy, as the infrastructure necessary for the support of IPM is often lacking and or is at a very elementary stage. Farm varsity based plant clinics may eliminate the role of middlemen with vested interests in the advisory system. Thus, onfarm adoption of IPM may occur only after certain degree of professionalism gains ground in our village economy.

IPM- Success Stories

Success stories of Integrated Pest Management where modules consisting of bio-agents, bio-pesticides, botanicals and minimum use of chemical insecticides like pre-sowing seed treatment with imidacloprid were used in crops like cotton, rice, groundnut, mustard, cabbage and citrus have been compiled by Vennila et al. (2016) in a joint publication of ICAR-National Research Centre for Integrated Pest Management. A full discussion of IPM in all above crops is beyond the scope of this book. To illustrate how IPM successes can lead to sustainable crop production, conservation and enhancement of natural enemy biodiversity, the salient features of IPM in cotton crop at Ashta village in Maharashtra, India, as meticulously compiled by Sharma et al. (2016) in the above cited ICAR compilation is discussed. Ashta village is part of a prime cotton growing region, has multiplicity of cultivars, staggard sowing dates, year round availability of American boll worm, *Helicoverpa armigera* and imbalanced use of fertilizers, which makes it ideal for this study. Farmers often use toxic pesticides like cypermethrin, dimethoate, fenvelerate, endosulfan etc. to manage various bollworm pests and sucking pests. Modules consisting of bio-agents, bio-pesticides and botanicals were evaluated. Seed treatment with imidacloprid was retained to keep off sucking pests during initial growth stages. A row of *Setaria* grass was planted after every 9-10 crop rows to provide live perches for predatory birds.

5.8.2 IPM-Strategies for Pest Management

The prime objective of various strategies is to minimize the use of toxic chemical pesticides, promote healthy environment, sustainability of crop protection measures with economic returns.

5.8.2.1 Farmer and Cultural Practices

Time tested mixed cropping systems where farmers grow locally adapted genotypes with diverse genetic base, not only help minimize risk of pest epidemics but also confer sustainability in our agroecosystems.

Most of such genotypes have disappeared after adoption of green revolution crops. Farmers' evolved cultural practices like adjustments in planting/harvest dates along with optimum crop diversity have sustained cultivation of food crops like rice, legumes, vegetables and even citrus crops under minimum input conditions in semi-arid environments of peninsular India and Africa. Early planting of chickpeas and mustard crops in the plains of northern India not only makes better use of depleting soil moistures, but also results in reduced pod borer and aphid infestations, as early maturation escapes the peak pest densities.

5.8.2.2 Host-Plant Resistance (HPR)

Brown plant hopper resistance in rice cultivars, stem borer resistance in maize and shoot fly resistance in sorghum and limited resistance to pod borers and diseases in chickpea and pigeonpea have helped reduce the overall pesticide load on the food crops. Such successes are often limited to local areas where resistant crop cultivars are identified. These often fail to be replicated over larger geographical areas and latitudes. The reasons are simple to explain; we identify resistant crop genotypes from amongst thousands of crop variants in rice or in chickpea crop as resistant to brown plant hopper, *Nilaparvata lugens*, and pod borer, *Helicoverpa armigera* respectively, but fail to appreciate that pest populations may also have thousands of variable pest populations across the geographical space and latitudes. It is not just a question of one or two biotypes, the biological species of insect pests are no longer considered as reproductively isolated populations. Variability is the result of frequent hybridization among variable pest populations, even sub species and related genera, mutations, genotype environment interactions etc. Despite all this, HPR still remains a most relevant strategy for pest management, because its three components, like non preference, tolerance and antibiosis still express, even though in variable amounts, across the geographical space.

Despite concerns about variability in expression of HPR, it is the most desired strategy in any crop improvement program as it aims to lessen total pesticide load through biological means. It has all along been the major strategy in the crop improvement program of the Indian Council of Agricultural Research (ICAR) and at ICRISAT, India. Sharma et al. (2005) have shown that the highly polyphagous insect, *Helicoverpa armigera* can discriminate between chickpea and pigeonpea varieties as evidenced by several resistant genotypes identified across the latitudes in India. They have suggested that resistance to pod borer should be given as much importance as yield to identify new varieties for cultivation by farmers. The incorporation of multiple disease resistance in improved cultivars of legume crops have been suggested (Nene, 1988). A similar multigenic HPR as strategy for management of insect pests would be more relevant. It seems relevant that with the help of advances in cell biology and biotechnology, crop improvement programs consider making cautious use of gene sequences from the widest possible genepool of the naturally evolved plant and animal species biodiversity. It is important that improved crop varieties are screened for host-plant resistance. Painter (1951) highlighted 3 basic types as '*non preference*', which may be due to various reasons; second '*antibiosis*', due to toxic effects in plants and the third '*tolerance*', a variety may tolerate larger insect damage without reduction in yield.

5.8.2.3 Biological Control

Agronomic practices aimed at conservation and augmentation of natural and endemic biological control agents like viruses, nematodes, parasitoids, predators and antagonists will always remain an important component of IPM.

There is general agreement that diverse assemblage of natural enemies like predators, parasitoids and pathogens are more effective in the control of herbivore pests (Denoth et al., 2002). There are instances when diverse natural enemy communities inhibit biocontrol, because they often attack each other (Letourneau et al., 2011). The potential of these species specific control measures need to be studied and demonstrated in on-farm situations. At present, the infrastructure and facilities for mass rearing and multiplication of biocontrol agents are available only at few places in India and other developing countries.

Field populations of the pod borer, *H. armigera* can also be controlled with applications of nuclear polyhedrosis virus. The use of HaNPV has emerged as an ecofriendly option for the management of *Helicoverpa* populations (Jayraj et al., 1989). Likewise, several strains of the bacterial insecticide, *Bacillus thuringiensis* (Bt), which produce an insecticidal crystal protein have been identified for the control of lepidopteran and dipteran pests. However, recent reports of resistance to this bacterial insecticide in bollworm and diamond back moth populations in Asia and Australia have caused concern among entomologists. Similarly, recent demonstrations of biological control of plant pathogenic fungi, by conserving and augmenting antagonists like *Trichoderma* spp. show the unlimited potential of biocontrol agents in IPM. Debjani and Farooqi (1989) compiled information on braconid parasitoids of *Helicoverpa*. Natural enemies, if conserved, have great potential for control of *Helicoverpa* (Greathead and Girling, 1989). The requisite infrastructure for mass rearing for augmented field releases still need to be developed.

Biological control being part of the overall strategy for natural biotic resource management is precisely the management of natural enemies like predators, parasitoids and pathogens of pests and of selected beneficial organisms like antagonists etc. and their ability to reduce pest populations below the economic injury levels. The need for biological control as an eco-friendly option in IPM gained tremendous appeal and popularity because of:

- Frequent crop failures due to ineffective pesticide.
- Growing concern over the harmful pesticide residues in food and water.
- Occurrence of wide spread pesticide resistant pest populations
- Adverse effects of pesticides on non-target organism's diversity
- Reduction of naturally occurring diversified biological control agent's populations.

Biological Control Agents' Biodiversity

Biodiversity resources including those of naturally occurring biological control agents include a rich diversity of parasitoids-trichogramatids, encyrtids, braconids, ichneumonids, tachinids etc; predators-like coccinellids, neuropteran lace wings, bugs, wasps, ants, spiders, reptiles, birds and mammals besides entomopathogenic fungi, viruses, protozoans, nematodes and antagonistic fungi. Large biodiversity of the naturally occurring biological control agents is still unexplored and fortunately, it still remains preserved meticulously in large number of natural refugia consisting of natural forest ecosystems and in the relatively undisturbed parts of the great Himalayan ranges.

Biological control has always been an important option worldwide. In India, research on biological control of harmful pest populations received impetus by the establishment of the Project Directorate of Biological Control (PDBC) in 1993 at Bangalore (now renamed as the National Bureau of Agriculturally Important Insects - NBAII) with broad based objectives for the management of insect pests, disease pathogens and weeds with the help of biocontrol agents.

Three Basic Components of Biological Control

Introduction

Introduction implies identification of exotic biological control agents, which have shown promise for the control of specific pest populations in different localities, for import through quarantine regulation for the purpose of multiplication and augmented releases for the control of specific pest population. During 1981, Herran and his team at IITA, Nigeria had successfully controlled mealy bug populations that had devastated the *Cassava* crop, the staple food of millions in Africa, with mass release of biological control agents, beneficial wasp and ladybug. Within three months, the mealy bug had mostly disappeared. Thus, livelihood of over 200 million people was restored and about twenty million lives were saved. Later, during his work at ICIPE, he advocated the agronomic push and pull method to deal with insect pests, weeds and low yields in maize and other crops, which the farmer could apply without accruing costs. This means that certain plants attract pest insects, while other plants would attract beneficial insects. This is now one of the classic models of agro-ecology, being adapted to many different crops and cropping systems. Similar observations were made while working with *Helicovera* infestations in chickpea fields at Pantnagar in northern India. It was observed that plants like *Chenopodium album*, a dominant weed was preferred over chickpea for oviposition (Bajpai and Sehgal, 1993). Similarly, *Vicia sativa* was a preferred host at Hisar in northern India (Chauhan, 1990).

Augmentation

This implies identification of locally adapted and endemic naturally occurring biological control agents, which have potential for the control of specific pest populations in an agro-ecosystem. Necessary infrastructure for mass multiplication on a truly large scale for a select group of biological control agents and the development of related technologies for harvesting, storage, transportation, release and recovery still need to be developed in India.

Although a large number of biological control agents have been identified with major crop pests - like 120 on *Helicoverpa armigera*, 42 on *Spodoptera litura*, 34 on *Bemisia tabaci*; 30 on *Chilo partellus* and 30 on *Plutella xylostella* on crucifers (Singh, 2001), probably several more have still to be discovered. Proper exploitation of such large and diverse range of biological control agents for the management of harmful pest populations will require understanding of tritrophic or even polytrophic inter specific interactions involving a wide range of semiochemicals which opens up a whole new areas for behavioural and chemical ecology for future investigations.

Conservation

This implies adopting cultural practices, cropping systems, companion and relay crops together with effective weed management for the sole purpose of conserving the diverse naturally occurring biological control agent's populations which in turn give natural biological control of the pest populations. Benefits of conserving flora in and around the agricultural cropping systems have been discussed in previous section on Introduction.

5.9 BOTANICAL PESTICIDES

Since prehistory and time immemorial, various plants and plant products have been used for crop protection and as medicine to treat several human ailments like infections, respiratory and stomach disorders.

5.9.1 Botanical Pesticides: Future of Pest Management

The plant kingdom, with more than 10,000 known secondary metabolites and expected diversity of more than 400,000 organic chemicals (Grainge and Ahmad, 1988) will always remain nature's biggest resource for sustainability of man and his agriculture.

Besides azadirachtin, nicotine, rotenone, pyrethrum etc. which protected our early agriculture, this chemical diversity also contains several steroids as insect hormone mimics which act as moulting inhibitors or accelerators, thus affecting insect growth. Almost 2,400 plant species having broad spectrum insecticidal properties have been found to control more than 800 species of insect pests (Grainge and Ahmad, 1988). This compilation had stimulated a fresh look at botanical pesticides. As time passes, this list is bound to grow many fold. A large variability of plant species are yet to be explored. It is necessary that such plants/ plant extracts are subjected to intensive research and systematic toxicological studies.

5.9.2 Neem: *Azadirachta indica* A. Juss. (Meliaceae)

The ancient Neem tree of Asia has been the source of natural medicinal extracts for thousands of years. The latinized name of neem, *Azadirachta indica*, is derived from Persian. Azad means "free"; dirakht means "tree" also referred to as the free tree of India. Neem is believed to have originated in Burma and the arid regions of the Indian sub-continent. During the last century, it was introduced in arid zones of Africa. Now it is grown in many Asian countries, in tropical regions of the New World, in several Caribbean and in some Mediterranean countries. During the last decade, neem has been introduced and planted on a large scale in Australia, in the Philippines and in the Plains of Arafat near Mecca in Saudi Arabia. Neem is also being planted on a large scale in southern provinces of China, in Brazil, in Peru and other tropical countries. Indian history documents farmers having purposefully tended to and harvested neem trees for their beneficial uses as far back as the Vedic period, 1500-600 B.C. Until recent history, neem has been used internally and externally in the traditional Ayurvedic herbal healing treatments in India.

The antiquity of knowledge about biomedical and biopesticidal uses of neem and other botanicals for human welfare has revealed fascinating information that dates 100,000 year in the prehistoric times and sheds light on rise and spread of Dravidian and human civilizations in India, Mexico, Africa and South America and China (Kumar and Navaratnam, 2013). Much of the following brief is an adaptation from their review.

Recognition of the potential of Neem for human welfare in America is recent. The National Academy of Sciences (1992) has described Neem as a tree for solving global problems. United Nations Environment Program (2012) has declared Neem: The UN's tree of the 21st Century.

Neem has been referred to as omnipotent tree, together with most of angiosperm flora and the vertebrate fauna had coevolved and diversified long before the continental split, when earth was a single supercontinent, called Pangaea that existed during the late Paleozoic and early Mesozoic eras. Siddha medicine practised then is one of the oldest medical systems known to mankind. Its origin goes back to 10,000 years B.C to 4,000 years B.C. Kumari Kandam civilization went back to 50,000 years B.C. In the annals of the ancient Siddha system of medicine, the first medicinal plant mentioned was neem or margosa. Knowledge about Siddha system of medicine was spread during Tamil Sangams. The first two Sangams took place in a sunken continent, Kumari Kandam. The first Sangam located in Thenmadurai, had 4,449 poets and 549 members including the father of Siddha medicine Agathiyar, also the head of the Sangam that lasted for 4,440 years (Wikipedia, 2012 e, f, g) (Fig. 16). Agathiyar specialized in medicine, language, alchemy and spirituality.

Neem use in Prehistory

Neem has been used by humankind from time immemorial as a deterrent for smallpox and other infectious diseases and is also considered to possess powers to ward off evil spirits. The earliest mention about the uses of medicinal plants were found in Thirumular Thirumantiram-Ennayiram, Tholkappiam and the ancient Tamil works of Sangam literature, which were believed to have been written thousands of years ago (I love India, 2012).

Fig. 16. *Agathiyar, the Head of Tamil Sangams that lasted for 4,440 years specialized in medicine, language, alchemy and spirituality.*

Spread through human civilization: Spread of Siddha mission to China

Bhogar Siddhar proceeded to China to spread the knowledge of Siddha sciences as per the last wishes of his guru Kalangi Nathar who belonged to China (Murugan: India. 2012). Kalangi Nathar instructed Bhogar in all aspects of the Siddha sciences, which included the preparation and use of the kaya kalpa herbal formulae to promote longevity. After Kalangi Nathar entered into Samadhi, Bhogar resumed his teaching mission to the Chinese. He transmigrated and thereafter went by the name "Bo-Yang" (Palani, India. 2012). Bo-Yang became also known as Lao-Tzu ("Old Master") and was accessible for nearly 200 years, and trained hundreds of

Chinese disciples. In the fifth century B.C, Confucius met Lao-Tzu and afterwards said that "*Today I have met Lao-Tzu, who is perhaps like the ascent of a Dragon on the wind into heaven is something which is beyond my knowledge*". Bo-Yang's leading disciple was Yu, whom he also gave the Indian name Pulipani. At the end of his mission to China, about 400 B.C, Bo-Yang with his disciple Yu and other close disciples, left China by the land route. As recorded in the Taoist literature, at the request of the gatekeeper at the Han Ku Mountain pass, Lao-Tzu crystallized his teachings. He did so in two books, the *Tao Ching*, with 37 verses, and the *Te Ching* with 42 verses. Upon Bhogar's return to Tamil Nadu, India, he submitted his 7,000 verse manuscript for evaluation to his guru, Agathiyar at Courtrallam and to an academy of Siddhas there. It was endorsed by all of them as a great work.

Later, after the period of the six Dynasties, 220 to 590 A.D, Bhogar returned with some Tamil disciples to China as a second mission to China. He left his mission in Tamil Nadu with Pulipani (Yu), the Chinese Siddha. As per recorded traditions of southern India, there are 18 siddhars who attained perfection, which included their spiritual, intellectual, mental, vital and physical bodies.

Neem in Siddha:

Potential in agriculture and human welfare

Knowledge about Siddha system of medicine and Neem has been documented and is available at the National Institute of Siddha, Chennai (2012), United Nations Educational, Scientific and Cultural Organization, Paris (2012) and at Wikipedia, San Francisco: Wikipedia (2012 h).

Typically Siddhars were saints, doctors, alchemists and mystics who wrote their findings in Tamil in the form of poems on palm leaves which are collected and stored in what are known today as palm leaf manuscripts in Tamil Nadu and handed down through the generations to public institutions, universities in India, Germany, Great Britain and USA.

5.9.3 Recent Biopesticidal uses of Neem

All parts of the neem tree have some medicinal value. Besides, neem has demonstrated a good potential for the management of major insect pests in rice, wheat, legumes and other agro-horticultural crops.

A review on "Management of insect pests with neem: Global perspective" by Saxena (2001) highlights biopesticide uses of neem.

These include potential of neem coated urea (Vyas et al., 1996), control of diamond back moth (Talekar and Shelton, 1993), mosquito repellent action of neem (Sharma et al., 1993), efficacy of neem for management of root-knot nematodes on tomato (Parveen and Alam, 1998) and neem seed derivatives for control of nematodes affecting banana (Musabyimana and Saxena, 1999). Later, neem seed kernel extracts have been extensively used in insect pest management in almost all of the crop improvement projects of the Indian Council of Agricultural Research.

5.10 GENETICS AND PLANT/ANIMAL BREEDING

Each living plant and animal species population has a phylogenetic lineage/ history of millions of years. Yet, each individual in a population holds only a small portion of the total gene pool of the species population for a short span of its life. Individuals die yet populations live on in a constant Darwinian struggle for survival of the fittest. Each species population of green plant or of animal species has evolved a specific niche to sustain life. Recent advances in the field of population improvement and quantitative genetics have shown that desired quantitative and qualitative traits/genes can be identified now more so with help of molecular genetics, gene mapping and bioinformatics even from a diverse range of both plant and animal species. The real issue is how beneficial traits can be identified and incorporated in the desired breeding program.

Agro-biodiversity should be the prime source for sustainable agricultural development, not just a source of traits in breeding programs (Conner and Mercer, 2007). The concept of sustainable agriculture could be a means of solving fundamental and applied issues related to food production in an ecological way (Lal, 2008).

5.10.1 Genetics and Plant Breeding

A variety is a constantly evolving dynamic natural plant population that are subject to hybridization and Darwinian natural selection. If left unmonitored, it will slip back to a wider genepool of the species population. Selected varieties with broad base MST's for multiple desired traits may have better chances of ecological sustainability.

The traditional approach in crop variety improvement has had major focus on productivity gains. The usual practice is to screen closely related cultivars with such traits and back cross with local genotypes and select from segregating F2 populations. Reasonably, homozygous populations giving enhanced yields are advanced further and purified through rouging off plants. Such varieties, as they are based only on Mendel's laws of inheritance of single morphogenetic trait are usually monogenic, often require constant monitoring and also fail sustainability in onfarm situations. Improved varieties with a multigenic broader base of high yield combined with resistance to biotic and abiotic stresses would have better prospects of sustainability on farmer's field.

Crop variety improvement has often been limited to hybridization between closely related varieties and aimed at incorporation of desired agronomic characters of higher yield with some stress tolerance. These efforts have met with limited success. With new techniques of cell biology and biotechnology, crop improvement should also consider hybridization between widely adapted varieties in its geographical range, closely related native species and even genera to identify and incorporate desired traits of host-plant resistance to major pest populations and to address problem of malnutrition in humans. Host-plant resistance and traits to fight malnutrition should receive focus equivalent to that on yield. This has also been stressed by Sharma et al. (2005) while reviewing host-plant resistance against *Helicoverpa armigera* (Hübner).

A variety or biotype is neither a biological species nor does it qualify to be a sub-species. At most, it may be considered as a *cline* if it occupies a wide range of geographic distribution or altitude variation, like the variations in conifer pines and several of the angiosperms across geographic areas and altitudes. A variety still holds only a small portion of the protected genepool of the species population, as long as it is monitored for homozygosity of the selected traits. Excessive homozygosity in any variety for a single trait selected for cultivation in large monocultures may not be ecologically sustainable. Selected varieties with broad base MST's for multiple desired traits may have better chances of ecological sustainability.

Based on '*Auxillary Taxonomic Principle*', Huxley (1939) used the term *cline* to indicate character gradient between same species populations. Such *clines* are more common in plants as has been documented for *Achillea* and *Pinus* (cf. Stebbins, 1963). Ecotypes in plants are almost same as subspecies (Stebbins, 1963) and interspecific and even intergeneric hybridization is more common in plants than in animals (Mayr, 1942). Several examples to show that many plant species have successfully evolved from such hybridizations have been documented by

Stebbins (1963). Several species of crucifer genus *Brassica* are known to occur as allopolyploids of basic species.

Despite limitations, inter-/intra-specific hybridization as has been going on in natural populations, should also be considered as a strategy to introduce characters in food crops that may be beneficial to fight several afflictions that people suffer due to malnutrition and weak immune systems. This can happen only when a larger emphasis is placed on explorations of plant biodiversity and variability in plant species populations across the geographic range of their distributions. A large biodiversity of plant species variability is still unexplored. It is necessary that plant taxonomy and biosystematics are given due importance.

5.11 GENETICALLY MODIFIED (GM) CROPS, BIOTECHNOLOGY

GM crops are defined as new varieties of crop species developed by molecular modification through the insertion of foreign genetic material. Present day genetic diversity of existing natural biodiversity resources and arable land need be conserved to produce healthy, nutritious food with minimum inputs and risks. To begin with, we must clear our mind-set that GM crops are necessary to secure food production in the coming decades. Estimates of living natural biodiversity resources suggest that we have discovered less than 2% of natural variability of plant, insect and animal wealth. The claim that biotechnology can create in short time, where nature takes millions of years is short sighted and lacks appreciation of the manner in which natural selection selects, out of million options, the best suited to propagate further. The variability of undiscovered genepool of DNA of plant and animal resources has immense potential to offer a solution to every need and malady mankind suffers. The best of biotechnology has not yet succeeded to create simple DNA or other biomolecules, the precursers of life on earth. Its claim to modify complex plant and animal systems in ecologically sustainable way has no scientific basis. It is necessary to introspect, why we have suddenly stopped explorations of plant and animal biodiversity and bioprospecting of their traditional knowledge among local people? Thus, the debate over GM crops *vs.* focus on agricultural biodiversity seems irrelevant. GM crops' technology applied to about 10 food and fibre crops to deliver inadequately tested products with irrational justification about their ecological sustainability with the stated objective to eliminate hunger through productivity gains. Are we really short of food or resources, means and knowledge to develop future genotypes the Nature's way based on yet poorly explored agricultural biodiversity, a product Nature and Natural Selection has evolved over millions of years? Even with existing productivity, both developed and under-developed world wastes more than 40% food between harvest and mouth. The technology of developing GM crops can well be deployed to refine plant based or animal based products that would help eliminate several diseases of weak immune system and other ailments mankind suffers.

Agricultural biodiversity is essential for a sustainable improvement in food and nutrition security (Frison et al., 2011). Jacobsen et al. (2013) have concluded that much of the research funding currently available for the development of GM crops would be much better spent in other research areas of plant science, e.g., nutrition, policy research, governance, and solutions close to local market conditions if the goal is to provide sufficient food for the world's growing population in a sustainable way.

Genetically modified (GM) crops contain foreign DNA with desired characters together with marker gene introduced by methods of genetic engineering. Two methods are used to introduce the foreign DNA into the cultured plant cells. The first involves the use of a plant pathogen, *Agrobacterium tumefaciens* which causes crown gall in many plant species. The bacterium is first disarmed to remove the plasmid gene that causes galls; the desired DNA is then integrated through the bacterium into the cultured cells. The second method uses 'gene gun' approach that fires gold particles that carry foreign DNA into plant cells. Some of these particles enter the plant cell wall and then enter its nucleus, thus integrating into the cultured cells. GM crops claim to give higher yields, better nutritional value, longer shelf life as well as resistance to drought and insect pests.

Ecological concerns due to safety and wider contamination of plant/ animal biodiversity by escape of foreign DNA must be addressed before GM crops are released. It would not be out of place to reemphasize that nature has evolved enough diversity and natural remedies to every malady for the entire life's biodiversity that includes man as well. Nature itself is the medical doctor for the entire biodiversity. Do GM crops give better option for a healthy life style? It is too early to answer this question as we have just discovered less than 2% of this vital natural biodiversity resource. Why should we look at the foreign genes and the bacterial pathogen to modify the DNA of the natural biologically diverse populations? This biodiversity itself provides for widest possible genepool resource for man and his growing populations.

Despite productivity gains, the development of GM crops has raised new legal, ethical and economic questions in agriculture reflecting their socio-economic and environmental impact. The official policy is ambivalent, promoting GM crops and at the same time promoting organic farming because of socio-economic and environmental considerations (Azadi and Ho, 2010). The main arguments of GM supporters are safe food security, improved food quality and extended shelf life as the reasons, which will benefit not only both consumers and farmers, but also the environment (Wisniewski et al., 2002). The main disadvantages of GM crops listed by Azadi and Ho (2010) include less quality of foods, antibiotic resistance, potential toxicity from GM foods, possible creation of allergenicity, unintentional gene transfer from GM to non-GM crops or to wild plants, possible creation of new viruses and toxins, limited access to seeds through patenting of GM food plants, threat to crop genetic diversity, religious/cultural/ethical concerns; concerns for lack of labeling, concerns of animal rights group, concerns of organic and traditional farmers etc.

Many scientists, Governments and corporates support GM crops. "*Food security in private hands is no food security at all*". It must be critically evaluated by broad based scientific body and consumers and environmentalists and, in fact all stakeholders. It is in this context that concerns expressed by Wisniewski et al. (2002) and Azadi and Ho (2010) must be evaluated.

Since two decades ago, when the GM crops were first introduced there have increasingly been hot debates on the applications of gene manipulation. Currently, the development of GM crop varieties has raised a wide range of new legal, ethical and economic questions in agriculture. There is a growing concern reflecting the socio-economic and environmental impacts of GM crops, which aims to criticize their value for farming systems. While organic crops are promoted as environment friendly products in developed countries, they have provoked great controversy in developing countries facing food security and a low agricultural productivity. On the one hand, farmers are encouraged to accept and implement GM crops because of their higher productivity, while on the other hand organic farming is encouraged because of socio-economic and environmental considerations. A crucial question is therefore, whether GM crops can co-exist with organic farming.

5.12 ORGANIC AGRICULTURE AND BIODIVERSITY

Organic farming makes a great relevance for the resource poor farmer in thickly populated countries when we consider socio-economic conditions. Thickly populated countries such as India, China, Africa and Latin America are largely an agrarian society where over 85% farmer's land holdings are just about one hectare and even less. These farmers and farm workers constitute over 60% of human population and are an extremely important natural biodiversity resource. Despite being resource poor, traditionally they have led a healthy life style and have shouldered a responsibility to feed even billions of people.

All this changed since 1960's, when monoculture HYV's of Green Revolution crops of wheat and rice, responsive to high input technologies of fertilizers, irrigation and pesticides replaced their traditionally diversified stress tolerant crop cultivars. These cultivars, evolved through centuries of natural selection were irretrievably lost as a precious natural resource. All this made him dependent on the mercy of local money lenders which forced him to lead a hand to mouth existence. A combined nexus between money lenders, local fertilizer and pesticide dealers, together with aggressive marketing of pesticide and farm input industries and even encouraged by corrupt practices of selective Government subsidies compounded the miseries of innocent farmers. He never understood the mathematics behind his exploitation. He soon lost the control of his own crop produce. Besides all this, erratic unpredictable monsoon rains often prompted drought or floods leading to recurrent crop failures. The major reason this climate shift was the deliberate destruction of biodiversity of prime natural forest reserves along mountain slopes, along entire Terai regions of Himalayas, along the catchment areas of major rivers etc. Massive encroachments of mangroves along coastal areas made weather even more unpredictable. Despite all this, developing countries never attempted any serious strategy like linking rivers, creating a large network of watersheds for harvesting and management of vast natural water resources. Frustrated with such exploitation, when farmers started suicides, the Governments and politicians' conscious of losing their constituency blamed Mother Nature.

5.12.1 Relevance of Organic Farming

Large tracts of mid and high hills in Himalayas and similar mountain areas in Europe, Africa etc. are already organic by default and have great many opportunities of growing agro-horticultural crops like citrus, stone fruits, off season vegetables and diversified legume crops. Besides, the crops grown in such areas have relatively longer growing season and have higher protein and nutritional content. Groundnuts of Champawat in mid hills of Kumaon Himalayas are famous for higher protein content.

Breeding approaches of trees for reforestation and of forage grasses and other plants for revegetation of uncultivated pastures and range land has been described by Stebbins (1963) as akin to Organic Evolution in the sense that the finished product cannot be provided for cultivation, rather it must compete under natural conditions for growth. This approach has lot of relevance for revegetation of southern slopes in low and mid hills of mountain ranges like the Himalayas.

5.13 PLANTATION CROPS AND HILL AGRICULTURE

Plantation crops like tea, coffee, banana, coconut palms etc. should preferably be grown as organic. These crops are a welcome alternative for revegetation of deforested areas; however, they do not replace the natural forests. Similarly, agro-forestry with selected genotypes may also increase biodiversity. Plantation of roadside shade trees of suitable species also enhances biodiversity.

As a result of rural urban migration in search of white collar jobs, over two thirds of human population in developing and even parts of developed world is now living in over-crowded cities and towns. The infrastructure in such habitations is bursting at seams. Even a mild earthquake can sometimes cause massive casualties. It is not possible to impose infrastructure in thickly populated areas where residential streets become the hub of branded shopping stores. Parts of several such cities need to be rebuilt with due attention to town and urban planning with ample green spaces and tree plantations to house the valuable human resource in habitats that may with stand seismic activity of 7 + magnitude. Roads must be paved with concrete walkways friendly to senior citizens and disabled.

Rejuvenation of hill horticulture, massive introduction and adaptation of internationally acceptable genotypes of apple, pear, apricot, plum,

Kiwi etc., together with relevant post-harvest handling and processing industry will help conserve biodiversity and bring prosperity. Similarly, promotion of dairying and animal farms, like goat and sheep and related processing industry as part of hill agriculture seem necessary.

5.14 SHOULD CHEMICAL PESTICIDES BE BANNED?

Should chemical pesticides be banned to promote organic farming? This question has come up in several forums on sustainable agriculture. Chemical pesticides are mostly xenobiotic in origin. First, insect populations like human or any other biological populations are a constantly evolving dynamic natural biodiversity resource. Insects respond to selection pressures due to application of chemical pesticides by developing resistance. As we apply same or more potent chemical pesticides repeatedly, the insecticide resistance also increases many times. Chemical pesticides will need to be phased out gradually. Fortunately, our experience with evolution of insecticide resistance and its management suggests that as we remove such selection pressures, the populations tend to return back to their susceptible status (Mehrotra and Phokela, 1992). Limited pesticide use may be retained for seed treatment where necessary or as life saving options. Pesticide use in standing crops should be minimised and gradually replaced with sustainable non-chemical pesticide options like neem. Pesticide use on cotton crop should be minimised and then phased out gradually.

When we prepare country wide biotic stress maps of the severity of insect population or its damage, we find that severity of attack or damage is restricted only to small areas where peak populations have perfect synchrony to crop maturity stages. The susceptible populations can easily be managed by suitable agronomic practices or with botanical pesticides.

5.14.1 Chemicals and the Pesticides Industry

Chemical pesticide industries may now debate for more focus on developing stable formulations of botanical pesticides. They may also diversify into improved seeds, effective seed treatment and organic farming etc. In the interest of saving further degradation of precious human resource and larger interest of saving natural biodiversity, it is suggested that the pesticide and chemical industry should initiate a broad based debate and redefine their role as partners of sustainable cropping systems. It is not just the question of promoting safer or more toxic pesticide options or largely unimplemented IPM agenda. Pesticides are biological poisons and cannot be classified as safe and unsafe options. Selective pesticides and growth hormone applications for seed treatment of cereal grains and pre-sowing treatment of potato and sugarcane seeds, plantation crops etc. may be retained in the interest of crop health and for initial growth stimulus only. Unwarranted applications of growth hormone like oxytocin, is injurious to health and should be banned. Pesticide use in short duration standing crops should be phased out gradually. In long duration crops like sugarcane and plantation crops, selective applications that would promote initial growth stimulus may be retained.

What if DDT and other chemical pesticides were not inducted for agriculture use since early 1950's? For several centuries, we had not only managed agricultural crops through natural and botanical means but have had also evolved stress tolerant cultivars in most crops. Chemical pesticide industries may now debate for more focus on developing stable formulations of botanical pesticides. Likewise, they may evolve better crop growth promotors, organic fertilizers, better microbial and biological control options.

5.15 CONSERVATION

Realising dangers of the destruction of our diversified natural resource base, the UN Conference on Environment and Development at Rio de Janeiro in June 1992 in its Agenda 21 unanimously adopted by over 150 countries placed unprecedented focus on Conservation of Biodiversity for Environmentally Sustainable Development. The recent UN Conference on Population and Environment 2015 has once again reaffirmed the above commitment.

Zisenis (2015) discussed the implementation of the newly established International Platform on Biodiversity and Ecosystem Services (IPBES) through briefly reviewing its creation and reflecting on an original idea to follow a bottom-up approach of a network of networks for supporting sustainable development. It reports on the outcomes of the Second Plenary Session of IPBES in Ankara, Turkey, in December 2013. Conclusions are drawn from the implementation in Europe of Local Agenda 21 and the River Basin Management Plans of the EU Water Framework. The International Platform on Biodiversity and Ecosystem Services (IPBES) was started with many ideas but a recurring theme was to create an organization that was an "Intergovernmental Panel on Climate Change (IPCC)" to make biodiversity issues better known and relevant in society to mainstream biodiversity in decision-making (Larigauderie and Mooney, 2010). Climate change is complex and requires coordinated global approach. The shortcomings of IPCC have now been fully addressed in Sustainable Developmental Goals 2015, SDG, under the UNEP.

After over sixty years of agricultural development in South-East Asia, effective institutional setup that caters to the study, exploration, inventories and conservation of almost over one million known species of plant and insect biodiversity resources is still lacking. In USA, USEPA and similar setups in developed countries and UNEP programs are known to address such concerns.

It is relevant to point out that insect pollination and selection pressure resulting due to insect attack has been one of the important causes of diversification of angiosperm flora which provide over 90% of food crops on the dining table. Besides, insect activity also provides honey, resins, perfumes and several industrial products. The need for an institutional set up in India, South-east Asia and rest of the developing countries similar to National Bureau for Plant Genetic Resources (NBPGR) in India has become a necessity. There should be a large central institutional set up on the lines of Smithsonian Institution in USA to promote taxonomy and systematics of natural biodiversity resources and house a large natural history museum for education and to promote awareness. It is expected that the knowledge thus accumulated about natural biodiversity resources, which made the planet earth their natural home since over 600 million years, might provide valuable clues for the sustainability of man as a supreme natural biodiversity resource.

5.16 HOLISTIC AGRICULTURE

The primary focus in holistic approach to agriculture should be plant health and not multiple prescription based use of excessive irrigation and chemical fertilizers together with destructive non target interventions of weedicides, pesticides, nematicides etc. during the short period of crop growth. Agroecosystem management is essentially an ecological strategy. This involves cultivation of crop genotypes, resistant or moderately resistant to biotic and abiotic stresses in sustainable diversified cropping systems. Management of natural biodiversity of weeds and other flora in and around the cropping system, use of renewable and not so easily renewable natural resources within the biological means of biocycling renewable natural resources are important components. From an agricultural stand point, conservation of biodiversity resources in and around cropping systems, together with adoption of diversified cropping systems can go a long way to confer sustainability (Letourneau et al., 2011). This involves adoption of ecologically sustainable cropping systems, use of renewable and not so easily renewable natural resources relevant to threat perception of abiotic and biotic stresses, within the biological means of biocycling renewable natural resources; suitable agronomic and cultural practices and manage weeds and other flora in and around the cropping system.

Every single member of crop plant is a living natural resource that has to be nurtured from seed to harvest, the nature's way. As such, it is a fragile natural resource, more so during early stages. Once the soil is prepared as a natural resource, free from pathogens and harmful insects and stocked with adequate moisture and nutrients, it is ready to support the crop agro-ecosystem. Crop plants may also need initial growth stimulus to establish the required plant stand for optimum growth and yields. This can be done the nature's way of using non chemical inputs of organic manures and vast array of botanical options. Thereafter, if excessive subsidies of irrigation, chemical fertilizers and pesticides are avoided, the crop will strengthen its own immune system and develop longer tap roots and stronger stems to support the desired plant architecture. A combined approach of development and selection of MST and HYV's, sustainable crop ecology, cultural practices and precision based use of natural resources together with better agronomic practices may be necessary to achieve this result. With adequate use of organic manures fortified with neem and other botanicals, it is possible to ensure crop canopy with minimum destructive interventions to sustain optimum yields at harvest.

5.16.1 Three Pillars of Sustainable Development: Environmental, Economic and Social

The UN conference in June 2012 on sustainable development while celebrating the 20th anniversary of the Earth Summit at Rio de Janerio, defined three pillars of sustainable development as economic, social and environmental.

Enabling environment to reap benefits of demographic dividend, distinct role for women in natural resource management, conserving biodiversity and forests are needed to realize increased economic growth and sustainable agriculture. International initiatives of UNEP and newly formed 17 SDG's (Sustainable Developmental Goals) 25, September 2015, approved by more than 150 heads of countries, is indeed an historic moment. This has set the agenda for next 15 years to achieve sustainable climate, environment, and agricultural production systems. The 17 SDG's fully integrates social, economic and environmental dimensions of Sustainable Development:

- End poverty in all its forms everywhere.
- End hunger, achieve food security and improved nutrition and promote sustainable agriculture.
- Ensure healthy lives and promote well- being for all at all ages.
- Ensure inclusive and quality education for all and promote life long learning.
- Achieve gender equality and empower all women and girls.
- Ensure access to water and sanitation for all.
- Ensure access to affordable, reliable, sustainable and modern energy for all.
- Promote inclusive and sustainable economic growth, employment and decent work for all.
- Build resilient infrastructure, promote sustainable industrialization and foster innovation.
- Reduce inequality within and among countries.
- Make cities inclusive, safe, resilient and sustainable.
- Ensure sustainable consumption and production.
- Take urgent action to combat climate change and its impacts.
- Conserve and sustainably use the oceans, seas and marine resources.
- Sustainably manage forests, combat desertification, halt and reverse land degradation, halt biodiversity loss.
- Promote just, peaceful and inclusive societies.
- Revitalize the global partnership for sustainable development.

These goals, though ambitious are achievable as present production systems across the world has enough resources to meet such obligations to human societies. We need political will, social awareness, equitable sharing of natural and renewable biodiversity resources and sustainable resource mobilization. The efforts of an independent monitoring agency under the leadership of UNEP program may be necessary.

Chapter 6

Biodiversity and Socio-Economics of Man

6.1 BIODIVERSITY AND SOCIO-ECONOMICS OF MAN

The anthropomorphic term, the Man and His World, refers to all gender and age variants of the species *Homo sapiens* in time and space. Evolved as a supreme natural resource, man has been a latest entry on biodiversity scenario of planet earth. Man differs from all other animals, as he has evolved further into a social human being. How much social he is today is the question? Through better intelligence he has transformed the environment and the space he occupies. In disregard to privileges of other fellow humans, he is a strong proponent of territoriality and defends his space with all his might. When this world is armed to teeth with nuclear weapons of mass destruction (WMD) and its leaders show little restraint in going to war at the slightest provocation, how long the inevitable can be prevented? As the largest consumer base of human population is gradualy shifting from America and Europe to Asia, the latter will probably be the arena for testing of latest version of WMD's. God bless! one shudders to think of consequences. Several such questions need serious thought. Can we afford to blink and say *que sera sera*: whatever will be, will be, the future's not ours to see, like in the lyrics of famous Doris Day of yesteryears?

As members of the vertebrate group of animals, human young ones still require intense parental care. Behaviour often plays a major role in survival. The human behaviour is largely acquired and very little of it is inherited. Thus, good and healthy environment during upbringing is essential to realise its optimum potential. Documented history of man is now about ten thousand years, thanks to the discovery of Tamil Patras in South Indian State of Tamilnadu, as have also been discussed earlier. Most of this period witnessed the beginning of major civilizations and an early adoption of agriculture.

Having achieved bipedal erect walking posture, man had almost stopped evolving further, except in social and intelligence domain. A bipedal and an erect posture also put severe limits on physical growth. It was anticipated that increased intelligence and brain activity will lead to increased skull size. This never happened. Rather, there has been an observed trend to achieve better brain to body ratio through optimization of body height due to limits imposed by axial skeleton and the life style changes.

6.1.1 Life Styles

Today's largely indulgent lifestyle affects body, mind and soul. With limits set by fragile frame of human body, it is the mind which plays the devil's sport. Unsustainable indulgent life styles together with dishonesty, cheating, robbing, food and living habits mostly dictated by human greed and an indulgent mind, has led to several afflictions like stress, depression, addictions, dementia etc. Such pursuits in various spheres of human activity like education, services, friendship, human relationships, politics, religion etc. may give temporary push over those working hard to achieve excellence. Eventually, it is the hard work, honesty and truth which pay the richest dividend, although it is true that honest people learn and sometimes lose from their experience. The cheaters and dishonest often become laggards and unable to face competition, they often come under depression. Many amongst them, despite having degrees and diplomas, often seek reservations over those who worked their way through honesty and hard work. Even if they succeed, they often fail miserably in performance and take to vices like alcohol, drugs, abuse of society. Such people seldom contribute genetically to next generation. They often become a social burden and even revolt, abuse the parents, elders and also take to society crimes. It is difficult to correct them, unless self realisation dawns upon them. Quality rehabs and psychotherapy, if available at affordable costs can play a major role. State support is often not enough to bring such people into mainstream. Depression thus becomes a major killer, followed by health damage due to addiction to such vices. They often become prey to killer diseases like cancer besides suffering from respiratory and cardiac ailments.

6.1.1.1 Mahatma Gandhi: His Commands for Healthy Life Style

After the Second World War, which claimed millions of human lives, Mahatma Gandhi, the great philosopher and thinker of those times, preached non-violence and self retraint to save mankind. It requires a very high degree of moral strength and courage to follow the law of self restraint as advocated by Gandhi (1928-1947). The first edition of his treatise 'Self- Restraint *vs* Self- Indulgence' was sold out in less than a week. In the preface to the third edition published in 1947, he says that self-restraint requires inexhaustible store of energy and patience to keep evil thoughts away. The best prescription he proposed to keep evil thoughts away was non-resistance. That means ignoring the existence of evil thoughts, by continuous preoccupation with the duties or services that keep you engaged - body, mind and soul. Evil thoughts and evil deeds are impossible when you are so absorbed in the work you do. In preface to the second edition, he says self control and not self indulgence should be the normal state in life. It was published at a time during the post war era when several nations were trying to emerge from the pangs of the war and colonial rule in several parts of Asia and Africa. There was widespread hunger and starvation. Feudal landlord sytem of farming and bonded labour still prevailed in several parts of Asia. International food aid was the rule, until early 1960's green revolution. Much of this has already been discussed in earlier sections.

In the present context, several of the issues discussed by Mahatma Gandhi have lost their relevance and we would fail on several parameters of Mahatma Gandhi. But, we must remember that there are very very few people of the stature, knowledge, vision and understanding of Mahatma Gandhi.

Moralistically, much of what Gandhi discussed may still be justified, but now when you need to earn your own space and livelihood with hard work and set up your own nuclear family, certain amount of self indulgence is bound to creep in. However, with one's own experience and due to society's fear, one must control his own limits of indulgence. Theater, cinemas and audio-visual aids are now also considered as learning aids. Social get togethers and wider interactions are also necessary to sharpen skills and succeed. His advice that we are all representatives of God - to serve all that lives and thus to express God's dignity and love is still relevant. Nature and the beautiful world of God's creation present many temptations for indulgence, but it also presents many challenges to discover and explore its mysteries. We know so little of the living biodiversity and it seems we are still scraping the surface of knowledge. The most important issue is conservation of the precious human resource in the face of stress and competition the working environment demands.

6.1.1.2 Yoga

In several parts of world, like in Latin America, Africa and Asia, hunger, poverty, deprivation as well as abuse of basic human rights prevail largely due to inequal opportunities and sharing of natural resources. Understanding the wide dimensions of Yoga may restore some pride, dignity and social well being of Man.

The word Yoga is a Vedic era term in ancient India, derived from Sanskrit word *yuj* which means to add, to join, to attach, to combine. In ancient Indian literature, the great saint Vyasa who wrote the first commentary on *Yoga Sutras* considered yoga to mean *Samadhi* or to concentrate. The ultimate goal of Yoga is *moksha* or liberation. Yoga, now considered as a means of self discipline to control body and mind, is a system of philosophy. Yoga now is a meditative means of discovering dysfunctional perception and cognition to overcome sufferings to achieve peace and salvation as referred to in ancient Indian text, Bhagavad Gita and Yogasutras texts (White, 2014). It also means the raising and expansion of consciousness from oneself to being with everyone and everything as discussed within Hindu Vedic literature and in the epic *Mahabharata* (White, 2011).

Recent studies to determine the effectiveness of Yoga as a complimentary intervention for cancer, schizophrenia, asthma and cardiac diseases (Smith and Pukell, 2009; Sharma and Haider, 2012; Innes and Bourguignon, 2005; Vancampfort et al. 2012) have shown mixed or inconclusive results. Cancer studies suggest none to unclear effectiveness while other sudies suggest that yoga may reduce risks and help patient's psychological healing process. The 193 member United Nations General Assembly on 11 Dec. 2016 approved by consensus a resolution establishing 21 June as "International Day of Yoga". The declaration of this day came after the call for the adoption of 21 June as International Day of Yoga by Indian Prime Minister Narendra Modi during his address to UN General Assembly on 27 September 2014. In suggesting 21 June as the International Day of Yoga, he had said that the date is the longest day of the year in the Northern Hemisphere and has special significance in many parts of the world.

Yoga came to the attention of an educated western public in the mid-19th century with the publication of N. C. Paul's published treatise on Yoga Philosophy in 1851.The first Hindu teacher to actively advocate and disseminate aspects of yoga to the western audience, Swami Vivekananda toured Europe and the United States in the 1890s. The reception which Swami Vivekananda received was built on the active interest of intellectuals in England and Europe. Due to influence of Swami Vivekananda, Yoga Sutras of Patanjali primarily refers to Ashtanga yoga, also called as "Raja yoga", "yoga of the kings," a term which originally referred to the ultimate, royal goal of yoga, which is usually samadhi. It incorporates metaphysics, ethical practices, systematic exercises and self-development techniques for body, mind and spirit. This has led to larger debate in recent years on the Art of Living.

6.1.2 Art of Living

Almost every one you know will give a big advice on the desired art of living. Several of the babas, self styled Guru's religious heads, together with their own band of faithful followers render big sermons on mass media about their version of the art of living. If you listen to each one of such versions, you are bound to be confused about what is good and what is bad in life.

Listed below are some of the repurcussions that affect human life due to our life styles and the art of living. Some of these may be due to our ignorance for which proper medical advice may be necessary. Much of the following is an adaptation from the article "A Natural Approach to Rising Health Epidemics" by Tina May who is an outspoken consumer health advocate, award-winning investigative journalist, internet activist and science lab director. Holding a degree from Stanford University in Biology, she is the lead writer for Health & Fitness at The Mandated.

6.1.2.1 Ageing

Ageing is an irreversible metabolic process, yet it is a common experience that we seem to age faster at certain times. Environmental pollution, injury, failure, stress, depression, feeling of loneliness, too much brooding over missed opportunities may sometime overtake your conscious mind. Unsustainable food habits, willful neglect of personal issues, addiction to alcohol, drugs etc. may seem to age you faster. The best way out is probably to talk about it to your friends, seniors, parents and medical professionals.

6.1.2.1 Hearing Loss, Blindness

Listening to high decible noise, even louder music may cause hearing loss. It is probably difficult to restore to normal, however hearing aids may help to some extent. Blindness or vision loss, if not congenital, or due to physical injury, could be corrected. Professional help would be necessary to restore normal vision. Vision loss may also occur due to addiction to alcohol, drugs or due to country side liquor etc. This may also damage pancreas, liver, heart etc., and may lead to diabeties and even cancer.

6.1.2.3 Diabetes, Obesity and Free Radicals

Diabetes affects an estimated 422 million adults worldwide. The number has quadrupled since 1980, now affecting 8.4% of the adult population. The most common form is Type 2 diabetes, accounting for 95% of all cases.

In Type 2 diabetes, the body can't use insulin the right way. As Type 2 progresses, the pancreas makes less and less insulin allowing sugar to build up in the blood stream. Fortunately, Type 2 diabetes is preventable and treatable. If ignored, diabetes can lead to life-threatening complications like heart disease, foot damage or amputation, nerve damage and even Alzheimer's. Diabetes, whatever be the cause, if not controlled with professional medical help can threaten life. Obesity and extra weight, affects our body's ability to handle stress. It affects our joints and puts extra pressure on heart to pump blood to all parts of body. It has emerged as a major killer of human life. Free radicals are there in pesticides, air pollutants, in food and even in the water we consume. Free radicals are unstable molecules that damage the body tissue by oxidation. These accumulate due to consumption of fried food, alcohol or smoking. They react with various compounds oxidising them. High level of oxidative stress affect every organ of the body, speed up ageing and may cause several illnesses and even cancer. You may fight free radical damage with antioxidants, which are some times referred to as 'free radical scavengers'. Antioxidants help neutralize and eradicate free radicals before they damage healthy and vital molecules.

6.1.3 Economic Costs of Willful Destruction of Natural Biodiversity Resource have been Colossal

Social and economic costs of deliberate destruction of natural biodiversity resources including of people by humans since 1940 onwards are colossal and far beyond any economic parameters of accountability. This destruction of natural human habitats is far more in thickly populated countries of Asia and is even more so in areas where people to land and forest ratio is high.

6.1.3.1 Social Implications

As humans, we are privileged to walk only a few steps in this glorious heritage of natural biodiversity resources. Due to unsustainable destruction of natural habitat and our life style, we have crossed all biological limits for renewability of natural resources used. Despite knowing this, there are politicians, dictators, sectoral groups who cause mass extermination of innocent people and even threaten the use Weapons of Mass Destruction (WMD). In any misadventure causing mass catastrophe, man being the weakest link in biodiversity will be its first casualty. The left over planet earth may not remain habitable by those who resort to such misadventure. The damage by greenhouse gases may burn large holes in the protective ozone layer, which may threaten all kinds of terrestrial Life. All social evils, like lack of respect for the system, addiction to drugs and alcohol, cheating, thefts, dacoities, extortions and even rape and murder are behavioural manifestations of perpetual disillusionment and frustrations of people gone astray. Such lack of vision in unemployed youth, 20-40 years is of great concern for sustainability of social system. Time for making amends is now; else our youth will soon enter older ages, when their productive energies will decline. We may learn from developed economies of the West where corporates would pay a dignified minimum wage based on minimum human needs of life, food and health security, besides contributions to old age security and pension.

6.2 SOCIAL WELFARE STATE: DREAM OR ACHIEVABLE REALITY?

The FAO Rome Declaration on World Food Security in 1996 defined its three basic dimensions as: availability, accessibility and utilization, with a focus on nutritional well being. Berry et al. (2015) concluded that sustainability should also be considered as a long-term time dimension in the assessment of food security. They stated that basic dimensions of food security as availability, accessibility and utilization are person centered, while sustainability has global dimensions. The concept of sustainable diets can also play a key role as a goal and a way of maintaining nutritional well being and health, while ensuring the sustainability for future food security.

Most Governments have done a lot to mitigate human suffering, it is still a long journey to reduce poverty, social and gender inequalities, provide adequate health care, equitable sharing of natural resources, quality education, eliminate pollution, pesticide residues, improve quality of air, water and living environments. The reason behind all above is poor quality of infrastructure, living habitats, destruction of trees and little conservation of local flora and fauna, lack of social and agroforestry and crop diversification, value addition in cropping systems etc. That much needed change must come now. The simple realization that basic human needs of all people are similar and the poor, irrespective of their work, must enjoy same respect and dignity. This does not mean there would be no differences in living standards of people. People who advance due to hard work, in professional and social domains must enjoy the status, dignity and rewards relevant to their contributions in society.

6.3 BASIC HUMAN RIGHTS: FOOD, NUTRITION, LIFE AND MEDICAL SUPPORT, SECURITY AND OLD AGE SUPPORT

This world has fairly good share of natural biodiversity resources in every region and people are working hard and have won laurels across the world. Yet developing countries houses the largest number, almost 80% of the world's poor and deprived people. They suffer malnutrition, poor health and inadequate health care. To compound their misery, millions do not even have a national identity. During any natural calamity, thousands perish even without having any identity. This is despite the fact that collective effort of almost 60% of its human resource consisting of small farmers and farm workers cultivate the rural land to produce food that feeds almost 7 billion people, yet many of those who produce food go hungry and remain undernourished. This happens even as almost 40% food produced in the world is lost between harvest and mouth. Right to life and medical support is basic fundamental human right, yet thousands of unaccounted people die and perish due to non availability of adequate medical support at affordable costs. Right to minimum habitat needs, like clean air, water and infrastructure and a decent old age security are a part of non-negotiable basic human rights. Ecologically sustainable living habitats are still beyond the reach of a vast majority of citizens.

6.4 SUGGESTED AGENDA

A large number of eminent economists and financial planners like *Niti Ayog* (Planning Commission) in India, and similar bodies across the world can best suggest equitable sharing of natural resources for human welfare and nation development.

6.4.1 A New Beginning

Our prime concern is dignified sustainability of precious human resource and the vast expanse of living biodiversity resources that support human survival. First, let us recognise every living individual, from new born to the oldest, irrespective of gender, cast, creed and status as equal member of the human resource and assign him undisputable identity. Given the present advances in computer technology, it can easily be done within a year. For general illness, this identity should entitle him to see any registered general physician in the country. This identity should also serve as a medical insurance number to seek the best possible medical care from reference to any registered medical specialist in the country for any critical illness. Consultation in every case must be reimbursed directly to the Doctor by the medical insurance through the human health welfare authority. Human resource development must ensure minimum basic employment of all unemployed youth, failing which they should be given career guidance and support for employable skills. All individuals, 18 years and above, should be given a social welfare number identity, linked to his/her unique identity. It should be mandatory for the employer to pay social welfare and health insurance, besides the minimum wage as per Government notification. Minimum wage even for unskilled workers must be adequate enough to support basic human needs for a family of three at market prices. For the period of unemployment, welfare authority must give support at minimum wage level or ensure new employment. Many welfare economies like those of USA, Canada, UK, etc. are implementing human social welfare programs, and we may learn and adapt some of the good features of them.

6.4.2 Empower Young People

Life styles may differ yet the basic minimum human needs of food, shelter, social, health and old age security, do not change with living status anywhere. This is same for every one whether a person is weeding a legume field in a cold winter morning on a farmer's field, or sitting in a cosy office to preside over the destiny of populations. People must earn their livelihood through a dignified wage commensurate with their skills, qualifications and status in the society and pay for health insurance and social welfare. Over 60% of young people in developing world do not get a dignified wage to meet these basic minimum human needs. It costs millions of dollars of tax payers' money to train skilled workers, engineers and doctors. Academic standards and education curricula must be improved to be at par with the Global Standards to prepare a critical human resource. Universities should not become commercial organizations to earn money without accountability to educational standards. Quite often, we encounter teachers, engineers and even doctors not earning a dignified wage that would satisfy their needs. Resource poor people have equal right of access to good quality, social, food and health security needs.

6.4.3 Nation Building an Uphill Task

Only sustained efforts, dignified wage structure, gender equality etc. can get the best from young minds and youth energies to develop world economies. When we expect the working classes to deliver the best of international quality products and services, why deny them the best of international wage structure and social welfare.Despite having world's largest share of poverty and malnutrition in developing countries, several million tons of food grains and agro-horticultural produce rot along road sides due to poor handling, storages and processing. Productive employment can easily be generated at local levels, which will direct youth energies into building social and physical infrastructure for the populations they share. It is time we start treating each member of human resource, irrespective of gender or any kind of social difference or classification, as a nation building unit and empower them with productive means that would generate far more resources than required to justify meeting basic minimum human needs of food, health and old age security. Let accessibility issues be decided by market forces of demand and supply, and guide the *Niti Ayog* to fix remunerative wage for productive work.

Chapter 7

Conclusions

7.1 CONCLUSIONS

The present day living biodiversity is the age of flowering plants, the Angiosperms, Insects, and Vertebrates including Man which are the most dominant and successful members of the terrestrial flora and fauna. These and the rest of consumer fauna have coevolved through reciprocal evolution ever since early Cretaceous almost 200 million years ago. Man the *Homo sapiens*, is the most recent arrival, since about 0.1 million years, as natural biodiversity resource of the vertebrate, bipedal primate lineage.

Marine biodiversity consisting of green algae, phyto-planktons and coastal mangrove flora have supported a rich diversity of consumer organisms consisting of sponges, jelly fishes, complex communities of corals, annelids, arthropods, molluscs, echinoderms, fishes and amphibians etc. over much longer period of 600 million years since Paleozoic era.

Biodiversity is not a static resource, rather it is dynamic, ever evolving, through evolutionary radiation, natural selection and Darwinian struggle of the survival of the fittest. This is a slow, gradual process of change spanning millions of years. This change, together with the changes in the geophysical environments, determines the phylogenetic relationships of the evolving lineages.

Another important lesson from bio-diversification spanning over 600 million years is that evolving geophysical changes had brought with it several phases of mass extinctions and of expansions of biodiversity's resources. Biodiversity has passed through several seral stages of ecological succession where community changes and their replacement by better evolved, more successful communities occur with regular frequencies. Despite all this, several of the major groups of plant and animal species which once enjoyed a dominant status have left behind legacy of few species which we call living fossils. We still have few representatives of Marcantiales, Filicales, Cycadales, Ginkoales and more in plant world. Several species of Xiphosura, the *Limulus*, Barnacles, several insects like Thysanurans, Proturans, Collemboles, even cockroaches, lace wings and Palaeopteran orders like Odonata, Ephemeroptera; several groups of molluscs like Nautiloidea, octopus, star fishes, Hemichordates and more in the animal world which still survive and occupy specialized ecological niches of their own. Such species form part of our glorious heritage, give glimpse of the past natural history.

By contrast, as individual members of this living world we hold only a small portion of the species gene pool. Events of last few decades since 1960's have witnessed massive destruction of natural biodiversity resources. During the recent memorable history of past few decades, people, due to man made and natural calamities like wars, sectorial violence, floods, sea storms etc, have killed more people. During this period, we have lost more natural forests than the present forest cover due to forest fires, systematic clearance of natural forests for farmland and urban needs. Man is now living far beyond the biological means of rejuvenation of lost natural biodiversity including that of man. Estimated rate of loss of natural biodiversity resources is almost 100 times more than the normal rate of species displacement. If unchecked, this together with next phase of mass extinctions due to natural or man made disasters of nuclear wars may bring exit of humans sooner than what would normally happen.

In early 1960's, adoption of high yielding Green Revolution varieties of wheat and rice in India and other Asian countries ushered an era of self sufficiency in food grain production. Within a span of few years by mid 1960's, the number of chemical fertilizer and pesticide industries increased from a single digit to over 500. Linking HYV's to high input applications of chemical insecticides was an ecological error. In unmonitored environments, people started consuming pesticide residues far above the maximum levels of WHO recommended limits. The incidence of cardiac, respiratory and diseases of the weak immune system also increased many fold.

Systematic destruction of million-year-old heritage of biodiversity of natural forests since early 1960's became the major cause of unsustainability of agriculture and human resource. Such events included massive clearance of natural forests, especially along the eco sensitive 100 km belt of *Terai* forests all along the foothills of Himalayas. Ecological disaster of this magnitude for handful gain of food grains caused irretrievable damage to the ecology and became a perpetual cause of human death and destruction due to floods and other natural calamities during the rainy season every year. This could have been avoided by partial conservation of sustainable forest stretches to conserve human settlements in a planned ecological way. Suitable orchards could have been developed to conserve the ecology of the region. This clearing of forests happened while vast stretches of northern plains with relatively low crop yields were available for expansion of farmlands.

Recent researches at local, national and international research institutions have demonstrated enough leads for safe, healthy and sustainable production system of agri-horticultural crops through collaborative partnership approach. Nations in thickly populated areas of world like India, China, etc. while they agree on agenda for conservation, exploration, inventorization and sustainable use of natural biodiversity resources have so far lacked political will and resources to accomplish this. As several wider issues of sustainability of human resource are at stake, multi national body, the United Nation must be directly involved not only in planning but also implementation of evolved agenda for ecological sustainability of well documented life's resources. There is a need for continuous debate and refinement of National Agriculture Policy, National Forest Policy and National Biodiversity Resources' Management Policy.

The following suggestions are just symbolic and not comprehensive and need be debated by enlightened body of stake holders.

- Initiate massive program for restoration, reforestation and rejuvenation of lost, depleted forests with suitable tree flora, wherever possible or put such areas for natural conservation for 100 years.
- Institutional set up at National, State and District Level to increase awareness of the relevance of natural biodiversity resources to sustainability of life's resources has become necessary.
- National Centres must be set up for conservation, exploration and utilization of natural biodiversity with special focus on dominant groups like plant biodiversity, insects and other arthropod groups etc. and their impact on sustainability of vital human resource, on the lines of Smithsonian Institutions of Natural History in USA.

- Education curriculum must be revised to give larger emphasis to Biological Science in applied disciplines of agriculture, horticulture, nutrition, medicine and even in social sciences.
- Large sections of remaining natural forests and mangroves must be brought under the ambit of National Parks with well planned policy for their conservation.
- Management of water as a natural resource for soil moisture recharge, and sustainable agriculture use seems necessary. Tapping of surplus water during rainy season, through large network of sustainable watersheds, sustainable linking and bunding of river water resources to prevent floods will help rejuvenation and conservation of natural biodiversity resources.
- In the interest of human health and to save small farmers from money lender debt traps, to prevent farmer suicides, a fresh National Agriculture Policy be evolved to delink adaptation of HYV's with costly inputs of large quantities of chemical fertilizers and scheduled applications of toxic pesticides.
- The above strategy can be achieved with combined use of organic manures, bio-fertilizers, botanical pesticides and microbial pesticides together with adjustments in cultural, agronomic practices and management of flora in and around cropping system to enhance conservation of natural biological control agents etc.
- Chemical pesticide use in standing crops need to be severely restricted and tuned to threat perception due to biotic and abiotic stresses.
- Minimum use of pesticides for seed treatment and presowing soil application in potato and long duration crops like sugarcane may be retained, if not disruptive to ecological sustainability.
- Broad based approach seems necessary to explore natural variability of genepools from sub species and from related genera to identify multiple genes for host plant resistance and multiple stress tolerance.
- Insecticide resistance management in crop pest populations will become much easier if selection pressures due to repeated applications of chemical pesticides is eased
- Promote organic agriculture in ecologically sensitive areas. Hill agriculture dominated by resource poor small farmers is largely organic by default. Certification and marketing of crop produce

as organic as per internationally accepted parameters will bring value addition.

- Rejuvenation of hill horticulture; massive introduction and adaptation of internationally acceptable genotypes of apple, pear, apricot, plumb, Kiwi etc., together with relevant post harvest handling and processing industry will help conserve biodiversity and bring prosperity.
- Promotion of dairying and animal farms like goat and sheep and related processing industry as part of hill agriculture seem necessary.
- Plantation crops like tea, coffea, banana and coconut may also be managed as organic for better returns.
- Relevance of GM crops and even Bt treated crops to sustainable human welfare and conservation of natural biodiversity resources must be established.
- Pesticide and chemical industry need to redefine their role as promoters of sustainable cropping systems.
- Pesticide industries may have more focus on developing stable formulations of botanical pesticides. Likewise, they may evolve crop growth promotors, organic fertilizers and tap large potential to develop and market improved seeds etc.
- For ages, botanical pesticides like azadirachtin, nicotine, rotenone, pyrethrum etc. have protected our early agriculture. This chemical diversity also contains several steroids as insect hormone mimics which act as moulting inhibitors or accelerators thus affecting insect growth. This option requires further research and development.
- As almost 98% of plant and animal species variability is still unexplored, tremendous wealth of secondary metabolites of plant and animal origin need to be explored for benefit of man and his agriculture.
- Lack of sustainable social welfare policy that gurantees basic human needs of non negotiable right to food, nutrition, health and social security entitlement have not only degraded young human resource, have also made several of them as addicts to alcohol, drug, tobacco and other vices.
- When almost 40% food produced by resource poor farmers and farm workers is lost between farm and mouth, such population

and the urban poor have non-negotiable rights to basic human needs of food, nutrition, health and social security.

- "Empower the Youth" to use energies in nation building tasks.

7.2 EPILOGUE

Increase in awareness of shrinking biodiversity and its impact on total natural biodiversity's resources including those of living humans the ruling classes and political leaderships will usher in a "True Social Welfare State" where every member can walk with his head held high and the mind without fear in true tribute to dreams of India's Nobel Laureate Rabindranath Tagore as below:

- Where the mind is without fear and the head is held high;
- Where knowledge is free;
- Where the world has not been broken up
- Into fragments by narrow domestic walls;
- Where words come out from the depth of truth;
- Where tireless striving stretches its arms towards perfection;
- Where the clear stream of reason has not lost its way
- Into the dreary desert sand of dead habit;
- Where the mind is led forward by thee
- Into ever-widening thought and action--
- Into that heaven of freedom,
- My Father, let my country awake.

Rabindranath Tagore

References

AICPIP (All India Coordinated Project on Improvement of Pulses). 1982-1988. Annual reports on *Kharif* and *Rabi* pulses. Entomology. 1981-82 to 1987-88. Directorate of Pulses Research, Kanpur, India.

Alfred, J.B.S., Das, A.K. and Sanyal, A.K. (eds.) 1998. Faunal Diversity in India. ENVIS Centre, Zoological Survey of India. Calcutta. pp.495.

Altieri, M.A. 1999. The ecological role of biodiversity in agroecosystems. *Agric. Ecosyst. Environ.* 74, 19-31.

Ananthakrishnan, T.N. and Sen, A. 1998. Biocommunication in Insects. Oxford & IBH Publishing Co. Pvt. Ltd., New Delhi. pp.104.

Azadi,H. and Ho, P. 2010. Genetically modified and organic crops in developing countries: A review of options available for food security. *Biotechnol. Adv.,* 28(1) : 160-168.

Bajpai, N.K. and Sehgal, V.K. 1994. Effect of neem, karanj and tobacco formulations on larval scrapping/ feeding responses on *Helicoverpa armigera* (Hubner). *Indian J. Appl. Ent.*, 8: 13-20.

Ball, G. E. 1979. Epilogue. In: *Carabid Beetles. Their Evolution, Natural History, and Classification.* (eds.Erwin, T.L., Ball, G.E., Whitehead, D.R., Halpern, A.L.) Dr. W. Junk Publishers, The Hague. pp 599-618.

Bardgett, R. D. and van der Putten, W. H. 2014. Below ground biodiversity and ecosystem functioning. *Nature*, 515 :505-511.

Berry, E.M., Demini, S., Burlingame, B., Meybeck, A. and Conforti, P. 2015. Food security and sustainability: can one exist without the other? *Public Health Nutr.*, 18(13):2293-2302.

Blackburn, T. M. and Gaston, K. J. 1996. A sideways look at patterns in species richness, or why there are so few species outside the tropics. *Biodiv. Lett.,*3, 44–53.

Blacket, P.M.S., Bullard, E. and Runcorn, S.K. (eds.) 1965. A Symposium on Continental Drift. *Philos.Trans. Royal Soc. A,* 258 (1088). The Royal Society of London. pp323.

Blade, S.F. 1998. Low-THC hemp production in Alberta. *Can. J. Plant Sci.,* 79:118.

Blade, S. F., Clayton, G.W. and Lyon, D. 2002 Crop diversification in the northern Great plains cropping systems. *Agron. J.*, 94: 173-74.

Brashares, J.S., Abrahms, B., Fiorella, K.J., Golden, C.D., Hojnowski, C.E., Marsh, R.A., McCauley, D.J., Nunez, T.A., Seto, K and Withey, L. 2014. Wildlife decline and social conflict. *Science*, 345: 376-378.

Brito, D., Moreira, D.O., Coutinho, B R. and Oprea, M. 2012. Ill nature: Disease hotspots as threats to biodiversity. *J. Nat. Conserv.*, 20(2):72-75.

Brooks, T.M., Akcakaya, H.R., Burges, N.D. et al. 2016. Analysing biodiversity and conservation knowledge products to support regional environmental assessments. *Sci. Data*. 3: 160007

Butchart, S.H.M., Walpole, M., Collen, B., van Strein, A. et al. 2010. Global biodiversity: Indicators of recent declines. *Science*, 328:1164–1168.

Cadotte, M. W., Cardinale, B. J. and Oakley, T. H. 2008. Evolutionary history and the effect of biodiversity on plant productivity. *Proc. Natl. Acad. Sci. USA.,* 105 : 17012–17017.

Cardinale, B. J., Wright, J.P., Cadotte, M.W., Caroll, I.T., Hector, A., Srivastava, D.S., Loreau, M. and Weis, J.J. 2007. Impacts of plant diversity on biomass production increase through time due to complementary resource use: A meta-analysis. *Proc. Natl. Acad. Sci. USA.,* 104 : 18123–18128.

Cardinale, B.J., Duffy, J.E., Gonzalez, A. et al. 2012. Biodiversity loss and its impact on humanity. *Nature*, 486 : 59-67.

Carson, R. 1962 Silent Spring, Houghton Miffin Co., Boston, USA.

Chapin, F. S., Zavaleta, E. S., Eviner, V. T., Naylor, R. L., Vitousek, P.M., Reynolds, H. L., Hooper, D.U. ,Lavorel, S., Sala, O. E., Hobbie, S. E. and Michelle C. 2000. Consequences of changing biodiversity. *Nature*,405 : 234- 242.

Chauhan, R.1990. Present status of *Heliothis* on pulses and strategies for its management in Haryana. In: *Helicoverpa management : Current status and future strategies.* (ed. J.N.Sachan). Directorate of Pulses Research, Kanpur, India. pp 50-54.

Chown, S. L., Clarke, A., Fraser, C. I., Cary, S. C., Moonand , K. L. and McGeoch, M.A. 2015. The changing form of Antarctic biodiversity. *Nature*, 522 : 431-438.

Chown, S. L., Gaston, K. J. and Williams, P. H. 1998. Global patterns in species richness of pelagic seabirds: the Procellariiformes. *Ecography* 21 : 342–350.

Ciegis, R., Ramarauskiene,J. and Martinkus,B. 2009. The concept of sustainable development and its use for sustainability scenarios. Inzinerine Ekonomika-Engg. Econ., 2 : 28-37.

Clarke, A. 2009. Temperature and marine macroecology. In : *Marine Macroecology* (eds. Witman, J. D. and Roy, K.). University of Chicago Press. pp. 250–278.

Crame, J. A. and Clarke, A. 1997. The historical component of marine taxonomic diversity gradients. In : *Marine Biodiversity: Patterns and Processes* (eds. Ormond, R. F. G., Gage, J. D. and Angel, M.V.). Cambridge Univ. Press, Cambridge. pp.122–147.

Cloud, P. E. 1969. Resources, population and the quality of life. Paper presented at AAAS meeting, Boston, MA. 29pp.

Conner ,A.J. and Mercer, C.F. 2007. Breeding for success: diversity in action. *Euphytica*, 154 : 261–262.

Corbett, J. 1954. Man-Eaters of Kumaon. Oxford University Press, New Delhi. pp.228.

Daily, G. C. 1997. Nature's Services: Societal Dependence on Natural Ecosystems. Island Press, pp. 412.

Dar, W.D. 2004. Foreword. In : *Heliothis /Helicoverpa Management. Emerging trends and strategies for future management*. (ed. H.C.Sharma). Oxford & IBH Publishing Co. Pvt. Ltd. pp v-vi.

Darwin, C. 1859. On the origin of species by means of natural selection. Jon Murray, London.

Debjani, D. and Farooqi, S.I. 1989. Biosystematic studies on hymenopterous parasitoids associated with *Helicoverpa armigera* (Hubner) in India - 1 (Family Braconidae). *Shaspa*, 6(2) : 105-113.

Decaens, T. 2010. Macroecological patterns in soil communities. *Glob. Ecol. Biogeogr.*, 19, 287–302.

Denoth, M., Frid, L. and Myers, J. H. 2002. Multiple agents in biological control : improving the odds? *Biol. Control*, 24 : 20–30.

Dethier, V.G. 1947. Chemical insect attractants and repellents. Blakiston, Philadelphia. pp. 289.

De Vries, H. 1909. The mutation theory : experiments and observations on the origin of species in the vegetable kingdom. Vol. I. The origin of species by mutation. English translation by Farmer, J.B. and Darbishire, A.D. Open Court Publishing Co., Chicago.

Dietz, R. S. 1961. Continent and Ocean Basin Evolution by Spreading of the Sea Floor. *Nature*,190 : 854–857.

Dirzo, R., Young, H.S., Guletti, M., Ceballos, G, Issac, N.J.B. and Collen, B. 2014. Defaunation in the Anthropocene. *Science*, 345 : 401-406.

Duffy, J. E. 2009. Why biodiversity is important to the functioning of real world ecosystems. *Front. Ecol. Environ*, 7 : 437–444.

Ehrlich, P.R. 2015. HT Report 21.06.2015. Stanford USA.

Ehrlich, P.R. and Raven, P.H. 1964. Butterflies and plants. A study in coevolution. *Evolution*, 18: 586-608.

Farrell, B.D. 1998. "Inordinate Fondness" Explained: Why are there so many beetles? *Science*, 281 : 555-59.

Fischlin, A. and Midgley, G.F. 2007. Ecosystems, their properties, goods, and services. In: *Climate Change 2007: Impacts, adaptation and vulnerability*. Contribution of Working Group II to the Fourth Assessment Report of the Intergovernmental Panel on Climate Change. (eds. Parry, M.L., Canziani, O.F., Palutikof, J.P., van der Linden, P.J. and Hanson, C.E.). Cambridge University Press, Cambridge. pp. 211-272.

Flather, C.H., Wilson, K.R., Dean, D.J. and McComb, B. 1997. Identifying gaps in Conservation Networks : of indicators and uncertainity in geographic based analyses. *Ecol. Appl.,* 7(2) : 531-542.

Flynn, D. F. B., Mirotchnick, N., Jain, M., Palmer, M. I. and Naeem, S. 2011. Functional and phylogenetic diversity as predictors of biodiversity-ecosystem-function relationships. *Ecology*, 92,1573-1581.

Fraenkel, G.S. and Gunn, D.L. 1940. The Orientation of animals. Dover Publications, New York. pp. 376.

Fraenkel, G. S. 1959. The raison d'etre of secondary plant substances. *Science*, 129: 1466-70.

Frick, K.E. 1959. Synopsis of species of Agromyzid leaf miners described from North America. *Proc.US Natl. Mus.*, 108: 347-465.

Friesen, G. 1987. The Canadian Prairies. University of Toronto Press, Toronto. pp. 543.

Frison, E, Cherfas, J and Hodgkin, T. 2011. Agricultural biodiversity is essential for a sustainable improvement in food and nutrition security. *Sustainability,* 3 : 238–253.

Gandhi, M.K. 1928-1947. Self Restraint vs Self Indulgence. Navajivan Publishing House. Ahmedabad

Gaston, K. J. 1996. Biodiversity- latitudinal gradients. *Prog. Phys. Geogr.*, 20 :466–476.

Gaston, K. J. 2000. Global patterns in biodiversity. *Nature*, 405 : 220-227.

Gaston, K. J. and Williams, P. H. 1996. Spatial patterns in taxonomic diversity. In : *Biodiversity: A Biology of Numbers and Difference* (ed. Gaston, K. J.). Blackwell Science, Oxford. pp. 202–229.

Gause, G.F. (1934). The struggle for existence. Williams and Wilkins, Baltimore.

Grainge M. and Ahmad, S. 1988. Handbook of Plants with Pest Control Property. John Wiley, New York. pp. 470.

Greathead, D.J. and Girling, D.J. 1989. Distribution and economic importance of *Heliothis* and of their natural enemies and host plants in southern and eastern Africa. In : *Proceedings of theWorkshop on Biological Control of Heliothis: Increasing the Effectiveness of Natural Enemies. New Delhi, India.* (eds. E.G. King and R.D. Jackson). Far Eastern Regional Office, U.S. Department of Agriculture, New Delhi. pp: 329-345.

Griffiths, G. C. D. 1967. Revision of *Phytomyza syngnesiae* group (Diptera: Agromyzidae) including species hitherto known as '*Phytomyza atricornis* Meigen'. *Stuttg. Beitr. Naturk.*, 177 : 1-28.

Griffiths, G. C.D.1974. Studies on boreal Agromyzidae (Diptera) V. On the genus *Chromatomyia* Hardy with revision of Caprifoliaceae mining species. *Quaest. ent.*, 10: 35-69.

Groombridge, B. (ed.). 1992. Global diversity status of world's living resources. London: Chapman and Hall. pp. 594.

Hansen, M. C., Potapov, P. V., Moore, R., Hancher, M., Turubanova, S. A., Tyukavina, A., Thau, D., Stehman, S. V., Goetz, S. J., Loveland, T. R., Kommareddy, A., Egorov, A., Chini, L., Justice, C. O., Townshend, J. R. G. 2013. High-resolution global maps of 21st-century forest cover change. *Science*, 342 : 850-853.

Hansen, M. C., Potapov, P. V., Moore, R., Hancher, M., Turubanova, S. A., Tyukavina, A., Thau, D., Stehman, S. V., Goetz, S. J., Loveland, T. R., Kommareddy, A., Egorov, A., Chini, L., Justice, C. O., Townshend, J. R. G. 2013. NASA-USGS Landsat Data Yield Best View to Date of Global Forest Losses, Gains. www.nasa.gov/content/goddard/nasa-usgs-landsat-data-yield-best-view-to-date-of-global-forest-losses-gains/

Hector, A. and Bagchi, R. 2007. Biodiversity and ecosystem multifunctionality. *Nature* 448, 188–190.

Hegerl, G. C., Hanlon, H. and Beierkuhnlein, C. 2011. Climate science: Elusive extremes. *Nature Geosci.*, 4 : 142–143.

Hennig, W.1950. Grungzuge einer Theorie der phylogenetischen Systematik. Deuscher Zentralverlag. Berlin. pp. 370.

Herren, H.R., Neuenschwander, P. 1991. Biological Control of Cassava Pests in Africa. *Annu. Rev. Entomol.*, 36: 257-283.

Hooker, Sir J. D. Flora of British India, 7 vol. (1872-1897) published by Drury (Heber).

Hooper, D. U., Chapin III, F. S., Ewel, J. J., Hector, A., Inchausti, P., Lavorel, S., Lawton, J. H., Lodge, D. M., Loreau, M., Naeem, S., Schmid, B., Setala, H., Symstad, A. J., Vandermeer, J. and Wardle, D. A. 2005. Effects of biodiversity on ecosystem functioning: A consensus of current knowledge. *Ecol. Monogr.*, 75, 3–35.

Hooper , D. U., Adair, E. C., Cardinale, B. J., Byrnes, J. E. K., Hungate, B. A., Matulich, K. L., Gonzalez, A., Duffy, J. E., Gamfeldt, L. and O'Connor, M. I. 2012. A global synthesis reveals biodiversity loss as a major driver of ecosystem change. *Nature*, 486 (7401) : 105-108.

Hopkins, A.D. 1919. The Bioclimatic Law as Applied to Entomological Research and Farm Practise. *The Scientific Monthly*, 8 (6) : 496-513.

Houghton, J. T., Jenkins, G. J. & Ephraums, J. J. (eds). 1990. Climate Change: The IPCC Scientific Assessment. Cambridge Univ. Press, Cambridge. pp. 410.

Huntingford, C. and Mercado, L. 2013. Beyond historical variability. *Nature*, 502 : 174.

Huxley, A. 1932. Brave New World. Chatto & Windus, UK. pp.163.

Huxley J.S. 1939. Clines: an auxillary method in taxonomy. *Bijdr. Dierk.*, 27: 491-520.

IAASTD. 2009. Agriculture at a crossroads. International assessment of agricultural knowledge, science and technology for development (IAASTD): Global Report. (Ed. BeverlyMcIntyre B. D. McIntyre, H. R. Herren, J. Wakhungu, R. T. Watson. pp. 606.

http://www.iloveindia.com/medicine-systems/siddha/. (Accessed on 12 August, 2012)

Innes, K. E. and Bourguignon, C. 2005. Risk Indices Associated with the Insulin Resistance Syndrome, Cardiovascular Disease and Possible Protection with Yoga: A Systematic Review". *J. Am. Board Fam. Pract.*, 18 (6): 491–519.

IPCC 2007. Climate Change 2007. Synthesis Report: Impacts, Adaptation and Vulnerability (Working Group II Contribution to the Fourth Assessment Report of the Intergovernmental Panel on Climate Change). p.12.

Isbell, F. Hector, A, Connolly, J., Harpole, W.S., Reich, P.B., Scherer-Lorenzen, M., Schmid, B. et al. 2011. High plant diversity is needed to maintain ecosystem services. *Nature*, 477 : 199–202.

IUCN. 2012. IUCN Red List of Threatened Species Version 2012.2. http://www.iucnredlist.org.

Jacobsen, Sven- Erik, Sørensen, M., Marcus, P. S. and Weiner, J. 2013. Feeding the world: genetically modified crops versus agricultural biodiversity. *Agron. Sustain. Dev.*, 33 (4) : 651-662

Jayraj, S, Rabindra, R.J. and Narayanan, K. 1989. Development of microbial agents for control of *Heliothis* in India. *In : Proceedings of the Workshop on Biological Control of Heliothis: Increasing the Effectiveness of Natural Enemies. New Delhi, India.* (eds. E.G. King and R.D. Jackson). Far Eastern Regional Office, U.S. Department of Agriculture. New Delhi. pp 483-504.

Jentsch, A., Kreyling, J. and Beierkuhnlein, C. 2007. A new generation of climate-change experiments: Events, not trends. *Front. Ecol. Environ.*, 5 : 365–374.

Johnson, K. G., Vogt, K. A., Clark, H. J., Schmitz, O. J. and Vogt, D. J.1996. Biodiversity and the productivity and stability of ecosystems. *Trends Ecol. Evol.*, 11 : 372-377.

Jorgensen, D. 2013. Ecological restoration in the convention on Biological Diversity targets. *Biodivers. Conserv.*, 22 : 2977-2982.

Joseph F. 2006. The lost civilization of Lemuria: the rise and fall of the world's oldest culture. Rochester: Inner Traditions/Bear &Co.pp.343.

Kerswell, A. P. 2006. Biodiversity patterns of benthic marine algae. *Ecology*, 87 :2479–2488.

Kevan, P.G., Clark, E. A. and Thomas, V. G. 1990. Insect pollinators and sustainable agriculture. *Am. J. Altern. Agric.*, 5 : 13-22.

Kormondy, E. J. 1996. Concepts of Ecology. Prentice Hall of India. New Delhi.559 pp.

Kremen, C. and Miles, A. 2012. Diversified farming systems: what is known about benefits, costs and tradeoffs among food production and ecosystem services? *Ecol Soc.*,17(4): 40.

Kumar, V.S. and Navratnam, V. 2013. Neem (*Azadirachta indica*) : prehistory to contemporary medicinal uses to humankind. *Asian Pac. J. Trop. Biomed.*, 3(7): 505-514.

Lal, R. 2008. Soils and sustainable agriculture. A review. *Agron Sustain. Dev.*, 28 : 57–64.

Lamarck, M. de. 1815. Histoire naturelle des animaux sans vertebres presentant caracteres generateaux et particulars de cesanimaux. Tome I. pp.1-462.

Larigauderie A. and Mooney H.A. 2010. The Intergovernmental science-policy Platform on Biodiversity and Ecosystem Services: moving a step closer to an IPCC-like mechanism for biodiversity. *Curr. Opin. Environ. Sustain.,* 2 : 9–14.

Larigauderie, A. Prieur-Richard, A.H., Mace, G.M., Lonsdale, M., Mooney, H.A., Brussaard, L., Cooper, D., Cramer, W., et. al. 2012. Biodiversity and ecosystem services science for a sustainable planet: The DIVERSITAS vision for 2012–20. *Curr. Opin. Environ. Sust.*, 4 : 101–105.

Latta, V.L.C., Baker, M., Crowl, T., Parnell, J.J., Weiner, B., de Wald, D.B. and Pfrender, M.E. 2011. Species and genotype diversity drive community and ecosystem properties in experimental microcosms. *Evol. Ecol.*, 25(5) : 1107-1125.

Lateef, S. and Reed. W. 1983. Review of crop losses caused by insect pests in pigeonpea, internationally and in India. In: *Crop Losses due to insect pests*. Special Issue. *Indian J. Ent.*, vol. II. pp. 284-291.

Letourneau, D.K., Armbrecht, I., Rivera, B.S., Lerma, J.M., Carmona, E.J., Daza, M.C., Escobar, S., Galindo, V., Gutiérrez, C., López, S.D., Mejía, J.L., Rangel, A.M., Rangel, J.H.,Rivera, L., Saavedra, C.A., Torres, A.M.

and Trujillo, A.R. 2011. Does plant diversity benefit agroecosystems? A synthetic review. *Ecol. Appl.*, 21 : 9–21.

Linnaeus, C. 1758. Systema Naturae. Regnum Animale (10th ed. Tomus 1; L. Salvii, Holminae).

Lotka, A.J. 1925. Elements of physical biology. Williams and Wilkins Co., Baltimore.

Loreau, M., Naeem, S. and Inchausti, P. 2002. Biodiversity and Ecosystem Functioning: Synthesis and Perspectives. Oxford University Press, New York. pp 294.

Lyons, S. K. and Willig, M. R. 1999. A hemispheric assessment of scale dependence in latitudinal gradients of species richness. *Ecology*, 80 : 2483–2491.

Mace, G. M., Gittleman, J. L. and Purvis, A. 2003. Preserving the tree of life. *Science*, 300 : 1707–1709.

Malcolm, J. R., Liu, C., Neilson, R. P., Hansen, L. and Hannah, L. 2006. Global warming and extinctions of endemic species from biodiversity hotspots. *Conserv. Biol.*, 20 : 538–548.

Malthus T. R. 1798. Essay on the Principles of Population. Harmondsworth. pp. 292

Mani, M.S. 1962. Introduction to High Altitude Entomology, Methuen & Co., London, pp 321.

Mani, M.S. 1964. Ecology of Plant Galls, Dr. W. Junk Publishers, The Hague. pp. 434.

Mani, M.S. 1968. Ecology and Biogeography of High Altitude Insects, Dr. W. Junk Publishers, The Hague. pp. 528.

Mani, M.S. 1974. Ecology and Biogeography in India, The Hague, Dr. W. Junk Publishers, pp. 7755

Mani, M.S., 1978. Ecology and Phytogeography of High Altitude Plants of the Northwest Himalaya. Chapman & Hall, London. pp. 205.

Matthews, G.A. 1984. Pest Management.Longmans, London. pp. 231.

May, R. 1988. How many species are there on earth? *Science,* 241 : 1441-1449.

Mayr, E. 1942. Systematics and the origin of species. New York. Columbia University Press. pp. 334.

Mayr, E., 1963. Animal species and evolution. The Belknap Press of Harvard University Press. Cambridge, Massachusetts. pp. 797.

Mehrotra, K.N. and Phokela, A. 1992. Pyrethroid resistance in *Helicoverpa armigera* (Hubner). Response of Populations in Punjab in Cotton. *Pestic. Res. J.*, 4: 59-61.

Mendel, G.J. 1866. Versuche über Pflanzen-Hybriden. In : Verhandlungen des natursforschenden Vereines in Brünn, IV. 3-47.

Mittermeier, R. A., Turner, W. R. Larsen, F. W., Brooks, T.M. and Gascon. C. 2011. Global Biodiversity Conservation: The Critical Role of Hotspots. In : *Biodiversity Hotspots*. (eds., Zachos F.E. and Habel J.C). Springer-Verlag, Berlin. pp. 3-22.

Mortimer, N., Campbell, H. J., Tulloch, A. J., King, P. R., Stagpoole, V., Wood, R. A., Rattenbury, M. S., Sutherland, R., Adams, C. J., Collot, J. and Seton, M. 2017. Zealandia: Earth's Hidden Continent, *GSA Today*, 27(3) : 27-35.

Myers, N. 1988. Threatened biotas: "hot spots" in tropical forests. *Environmentalist*, 8 : 187–208.

Myers, N. 1990. The biodiversity challenge: expanded hot-spots analysis. *Environmentalist*, 10 : 243–256.

Mora, C., Frazier, A. G., Longman, R. J., Dacks, R. S., Walton ,M. M., Tong, E. J. et al. .2013. The timing of climate change. *Nature*, 502 : 183–87.

Munday, P. L., Jones, G. P., Pratchett, M. S. and Williams, A. J. 2008. Climate change and the future for coral reef fishes. *Fish Fish.,* 9 : 261–285.

Musabyimana, T. and Saxena, R. 1999. Efficacy of neem seed derivatives against nematode affecting banana. *Phytoparasitica*, 27 : 43-49.

NASA/GODDARD SPACE FLIGHT CENTER PUBLIC RELEASE: 15-NOV-2013. NASA-USGS landsat data yield best view to date of global forest losses, gains. http://www.nasa.gov/goddard

National Academy of Science. 1992. Neem, a tree for solving global problems Washington, DC.

National Institute of Siddha. Chennai. 2012. National Institute of Siddha; About Siddha. http://www.nischennai.org//about- siddha.htm.

Near, T. J. Dornburg, A., Kuhn, K. L., Eastman, J. T., Pennington, J.N., Patarnello, T., Zane, L., Fernández, D. A. and Jone, C.D. 2012. Ancient climate change, antifreeze, and the evolutionary diversification of Antarctic fishes. *Proc. Natl. Acad. Sci. USA*, 109 : 3434–3439.

Neem Foundation. Mumbai: 2012. All about neem. http://www.neemfoundation.org.

Nene, Y.L. 1988. Multiple Disease Resistance in Grain Legumes. Ann. Rev. Phytopathol., 26 : 203-17.

Occasional Paper Series 3. 2008 Climate Change: Challenges to Sustainable Development in India. pp 28 Research Unit (LARRDIS) Rajya Sabha Secretariat, New Delhi.

Odum, Eugene P. 1971. Fundamentals of Ecology. 3rd Edition. W. B. Saunders Company.574 pp.

Parveen, G. and Alam, M.M. 1999. Efficacy of neem products for the management of root-knot nematode on tomato in soils polluted with heavy metals, cadmium and lead. In: *Azadirachta indica* A. and Juss. (eds. Singh, RP and Saxena, R.C.). Science Publishers, USA. pp: 235-244.

Painter, R.H. 1951. Insect resistance in crop plants. The Macmillan Company, New York. pp. 520.

Pimbert, M.P. and Sehgal, V. K. 1989. The relative importance of *Helicoverpa armigera* (Hubner) in different agro- ecological zones of India: Implications for Pest Management. Legume Entomology Discussion Paper. 1. ICRISAT, Patancheru, A.P., India. p. 29.

Pimbert, M.P., Lateef, S.S., Sehgal, V.K. and Wightman, J.A. 1989. *Helicoverpa armigera,* the worst pest problem in the Semi-Arid Tropics. In: *Integrated Pest Management in Tropical and Subtropical Cropping Systems*. Proceedings (Vol.1). Eschborn, Germany. Deutsche Gesellschaft fur Technische Zussamenarbeit. pp. 219-233.

Pimm, S. L., Russell, G. J., Gittleman, J. L. and Brooks, T. M. 1995. The future of biodiversity. *Science*, 269 : 347–350.

Platnick, N. I. 1991. Patterns of biodiversity: tropical vs temperate. *J. Nat. Hist*, 25 : 1083–1088.

Post, E. 2013. Tropical diversity countdown. *Nature*, 502 : 175.

Raven, P.H. and Wilson, E.O. 1992. A fifty year plan for biodiversity studies. *Science*, 258 : 1099-100.

Reed, W., Cardona, C., Sithanantham, S. and Lateef, S.S. 1987. Chickpea pests and their control. In : *The Chickpea*. (eds. M.C. Saxena and K.B. Singh). Wallingford, Oxon, UK. CAB International. pp. 283-318.

Richards, P.W. 1973. The Tropical Rainforest. *Sci. Am.*, 229(6) : 59-67.

Rhode, K. 1992. Latitudinal gradients in species diversity: the search for the primary cause. *Oikos,* 65 : 514–527.

Roy, K., Jablonski, D., Valentine, J. W. and Rosenberg, G. 1998. Marine latitudinal diversity gradients: tests of causal hypotheses. *Proc. Natl Acad. Sci. USA,* 95: 3699–3702.

Saxena, M.C. 1987. Agronomy of chickpea. In. *The Chickpea.* (eds. M.C. Saxena and K.B. Singh). Wallingford, Oxon, UK. CAB International. pp. 207-232.

Saxena, M.C. 1990. Problem and potential of chickpea production in the nineties. In : *Chickpea in the Nineties. Proc. of 2nd International Workshop on Chickpea Improvement.* ICRISAT Center, Patancheru. India. pp. 13-26.

Saxena, M.C. 2015. Innovative Technologies of Dry Land Agriculture for Global Food Security. *In* : *Awakening for Advancement of Science and & Technology. Souvenir cum Newsletter. Alumni Almamater Meet, 2015.* G.B. Pant University of Agriculture & Technology. Pantnagar, India. pp. 43–44.

Saxena, R.C. 2001. Management of Insect Pests with Neem: Global Perspective. In: *ICAR-IARI Summer School 'Alternate Strategies for Insect Pest Management in Major Crops.* 11 June – 02 July, 2001. Division of Entomology, Indian Agricultural Research Institute. New Delhi. pp. 160- 176.

Schoonhoven, L.M. 1972. Secondary plant substances and insects. *Rec. Adv. Phytochem.*,5: 197-224.

Schulze, E. D. and Mooney, H. A. 1993. Biodiversity and Ecosystem Function Springer-Verlag, Berlin. pp. 525.

Seddon, P. J, Griffiths, Christine J, Soorae, Pritpal S, and Armstrong, D.P. 2014. Reversing defaunation: Restoring species in a changing world. *Science,* 345 : 406-412.

Sehgal, V.K. 1965. Studies on Indian Agromyzidae (Diptera) – 2. *Beitr. Ent*, 15 (1/2): 3-10, Berlin.

Sehgal, V.K. 1971(a). Biology and Host-plant Relationships of an Oligophagous Leaf miner, *Phytomyza matricariae* Hendel (Diptera: Agromyzidae). *Quaest. Ent.*, 7: 255-280.

Sehgal, V.K. 1971(b). A taxonomic survey of the Agromyzidae (Diptera) of Alberta, Canada, with observations on host-plant relationships. *Quaest. ent.*, 7: 291-405.

Sehgal, V.K. 1990(a). Damage/ yield relationship due to *Helicoverpa armigera* (Hb.) larvae in chickpea in India. *In: Proc. of the Second International Workshop on Chickpea Improvement.* ICRISAT Centre, Patancheru. India.

Sehgal, V.K. 1990(b). Agro-ecological pest management in chickpea and pigeonpea crops in India. *In : 3rd International Conference on Plant Protection in the Tropics*. MAPPS, Genting Highlands, Panang, Malaysia.

Sehgal, V.K. 1999. Insect Biosystematics and Sustainability in Agriculture Tenth Dr. C.P. Alexander Memorial Lecture.19 Feb 1999. Department of Zoology, University of Delhi, Delhi, India. pp.16.

Sehgal, V.K. 2001. Principles and Practices in Insect Pest Management.. In. *ICAR-IARI Summer School 'Alternate Strategies for Insect Pest Management in Major Crops.* 11 June – 02 July, 001. Division of Entomology, Indian Agricultural Research Institute. New Delhi. pp. 5-10.

Sehgal, V.K. 2005. Biodiversity and *Heliothis/ Helicoverpa*Management. In *Heliothis/ Helicoverpa Management.* (ed. Sharma, H.C.). Oxford and IBH Publishing Co. Pvt. Ltd. New Delhi.

Sehgal, V.K. and Ram Ujagir, 1987. Plants' natural defenses in Cruciferae and Tropaeolaceae against mustard sawfly, *Athalia proxima* Klug. *Indian J. Ecol.*, 4: 199-205.

Sehgal, V.K., Rameshwer Rao, V. and Wightman, J.A. 1990. Relationship between pod borer density and yield loss in chickpea: One Season's Data. ICN. International chickpea newsletter. pp. 29-30.

Sen, A. and Sehgal, V.K. 1993. Resistance in Peas, *Pisum sativum* L., against pea leaf miner, *Chromatomyia horticola* (Goureau) (Diptera: Agromyzidae): Feeding and Ovipositional Preferences. *Insect Sci. Applic.*, 14(2) : 185-191.

Sen, A. 1998. Neuroethological approaches in insect-plant interactions. In *Biocommunication in insects*. (eds. Ananthakrishnan, T.N. and Sen,A.). Oxford and IBH, New Delhi. pp 54-68.

Sharma, M. and Haider, T. 2012. Yoga as an Alternative and Complementary Treatment for Asthma: A Systematic Review. *J. Evid. Based Complementary Altern. Med.*, 17 (3) : 212–217.

Sharma, H.C., Ahmad, R., Ram Ujagir, Yadav, R.P., Singh R. and Ridsdill-Smith, T.J. 2005. Host-Plant Resistance to Cotton Bollworm/ Legume Pod Borer, *Heliothis/ Helicoverpa.* In : *Heliothis/ Helicoverpa Management.* (ed. Sharma, H.C.). Oxford and IBH Publishing Co. Pvt. Ltd. New Delhi. pp. 167-208.

Sharma, O.P., Murthy, K.S., Puri, S.N., Lavekar, R.C. and Mayee, C.D. 2016. Integrated Pest Management in Cotton at Ashta Village. In : (eds.Vennila et al.,). *Success Stories of Integrated Pest Management in India.* ICAR-National Research Centre for Integrated Pest Management, New Delhi.

Sharma, V.P., Ansari, M.A. and Razdan, R.K. 1993. Mosquito repellent action of Neem (*Azadirachta indica*) oil. *J.Am. Mosq. Cont. Assoc.,* 9(3) : 359-360.

Shillong Times, 22/07/2016. Assam Govt. Report, 239 Rhinos killed in 16 years. Poaching, Wildlife Trade Endangered Species.

Siddiqui, I.A. 2015. WTO Talks and Agricultural Trade and Development. *In : Awakening for Advancement of Science & Technology.* Souvenir cum Newsletter. Alumni Almamater Meet, 2015. G.B. Pant University of Agriculture &Technology. Pantnagar, India. pp.45-47.

Simpson, G.G. and Beck, W.S. 1965. Life, an Introduction to Biology. Harcourt, Brace & World, New York.

Singh, N.K. 2007. The Politics of Change: A Ringside View. Penguin India. New Delhi. pp.288.

Sithanantham, S., Rameshwer Rao. V. and Gaffar, M. A. 1984. Review of crop losses caused by insect pests on chickpea. In : *Crop Losses due to insect pests.* Special Issue. Indian J. Ent., The Entomol. Soc. of India. Hyderabad. vol. II. pp. 269- 283.

Sithanantham, S., Sehgal, V.K. 1985. A survey for off- season survival of Pigeonpea pod fly around Pantnagar, India. *International Pigeonpea Newsletter*, 4: 53- 54.

Sneth P. H. A. and Sokal R. R. 1973. Numerical taxonomy: the principles and practice of numerical classification. Freeman and Company, San Francisco: 962 pp.

Sodhi, N. S. and Ehrlich, P. R. (eds). 2010. Conservation Biology for All. Oxford University Press, Oxford, UK.

Spence, J.R. 1979. Riparian Carabid guilds – a spontaneous question generator. In : *Carabid beetles.* (eds. Erwin, T.l., Ball, G.E., Whitehead, D.R. and Halpern, A.L.). Springer, Dordrecht.

Smith, K.B. and Pukell, C.F. 2009. An evidence-based review of Yoga as a complementary intervention for patients with cancer. *Psychooncology*, 18(5) : 465-475.

Spencer, K.A.1973. Agromyzidae (Diptera) of Economic Importance. Dr. W. Junk B.V.- The Hague. pp. 416.

Spencer, K.A. 1990. Host Specializations in World Agromyzidae (Diptera). Kluwer Academic Publishers. Dordrecht, Boston, London. pp. 444.

Stebbins, G.L. 1963. Variation and evolution in plants. Columbia University Press. New York and London. pp. 643.

Stevens, G.C. 1989. The latitudinal gradient in geographical range : how so many species coexist in the tropics. *Am.Nat.*, 133 : 240-256.

Stokstad, E. 2014. The Empty Forest. *Science*, 345 : 396-399.

Swaminathan, M.S. 1988. Our Common Agricultural Future. International Agriculture Newsletter. 125. pp. 1-2.

Swaminathan, M.S. 1994. Population, Environment and Food Security. Lecture delivered at International Conference on Population and Development, Cairo, Egypt, 9 Sept. 94.

Talekar, N.S. and Shelton, A.M. 1993. Biology, ecology and management of the diamondback moth. *Annu. Rev. Entomol.*, 38 : 275-301.

Tittensor, D. P. Mora, C., Jetz, W., Lotze, H.K., Ricard, D., Berghe, E.V. and Worm, B. 2010. Global patterns and predictors of marine biodiversity across taxa. *Nature*, 466 : 1098–1101.

TOI,15.08.2015. Uttarakhand Government citing Forest fire risk, plan to cut chir pine trees. Times of India report.

UNEP, 2012. Avoiding Future Famines: Strengthening the Ecological Foundation of Food Security through Sustainable Food Systems. United Nations Environment Programme (UNEP), Nairobi, Kenya.

United Nations Environment Programme Neem: The UN's tree of the 21st Century. Nairobi: United Nations Environment Programme; 2012

UNESCO, 2009. World Water Assessment Programme. Water in Changing World. The United Nations World Water Development Report 3.United Nations Educational. Scientific, and Cultural Organization, Paris:

UNESCO, 2012. The I.A.S. Tamil Medical Manuscript Collection. Available from: http://www.unesco.org/new/en/communication- and-information/flagship-project-activities/memory-of-the-world/register/full-list-of-registered-heritage/registered-heritage-page-8/the-ias-tamil-medical-manuscript-collection.

Vancampfort, D.,Vansteeland, K., Scheewe, T., Probst, M., Knapen, J., De Herdt, A. De Hert, M. 2012. Yoga in schizophrenia: a systematic review of randomised controlled trials. *Acta. Psychiatr. Scand.*. 126 (1) : 12–20.

Venail, P. A., Maclean, R. C., Meynard, C. N. and Mouquet, N. 2010. Dispersal scales up the biodiversity-productivity relationship in an experimental source-sink metacommunity. *Proc. R. Soc. Lond.* B , 277 : 2339–2345.

Vennila, S., Birah, A., Kanwar, V. and Chattopadhyay, C. 2016. Success Stories of Integrated Pest Management in India. ICAR-National Research Centre for Integrated Pest Management, New Delhi- 110 012. pp. 78.

Vignieri, S, Sugden, A.M. and Pennisi, E. 1914. Introduction to Special Issue Vanising Fauna. *Science,* 345 :392-395.

Visconti, P., Bakeenes, M., Baisero, D., Brooks, T., Butchart, S.H.M., Joppa, L. et.al. 2015. Projecting Global Biodiversity Indicators under Future Development Scenarios *Conserv. Lett.*, 9 : 5-13.

Volterra, V. (1926). Viariazioni e fluttuzioni del numero d' individui in specie animali conviventi. *Mem. R. Acad. Naz. Dei. Lincei.* Ser VI. 2.

Vörösmarty, C.J., et al., 2005. in Millennium Ecosystem Assessment Vol.1, Ch. 7, 165–207 (Island Press).

Vyas, B.N., Godrej, N.B. and Mistry, K.B. 1991. Development and evaluation of neem extract as a coating for urea fertilizer. *Fert. News.*, 36(2) : 19-25.

Vitousek, P.M. and Hooper, D.U. 1993. Biological diversity and terrestrial ecosystem biogeochemistry. In : Biochemistry and Ecosystem function. (eds. Schulz, E.-D. and Mooney, H.A.). Springer, Berlin. pp.3-14.

Wallace, A.R. 1876. The Geographical Distribution of Animals, with a study of the relations of living and extinct faunas as elucidating the past changes of the Earth's surface. Vol. I. Harper and Brothers. New York.

Wardle, D. A. 2002. Communities and Ecosystems: Linking the Aboveground and Belowground Components. Princeton Univ. Press. Wernberg.

Wegener, A. 1912. Die Herausbildung der Grossformen der Erdrinde (Kontinente und Ozeane), auf geophysikalischer Grundlage. *Petermanns Geogr. Mitt.,* 63: 185–195, 253–256, 305–309.

Wernberg, T., Small, T.A., Tuya, F., Thomsen, M.S., Langlois, T.J., de Bettigneis, T., Bennet, S. and Rousseaux, C.S. 2012. An extreme climatic event alters marine ecosystem structure in a global biodiversity hotspot. *Nat. Clim.Change*, 3 : 78-82.

White, D.G. 2011. Yoga, - Brief history of an idea. Princeton University Press.

White, D.G. 2014. The Yoga Sutra of Patanjali: A Biography. Princeton University Press.

Whitehead, D.R. 1979. Late glacial and post glacial vegetational history of the Bershires, western Massachusetts. *Quart. Res.*, 12 ; 33-357.

Whittaker, R.H. and P.P. Feeny. 1971. Allelochemics: Chemical Interactions between species. *Science.* 171. : 757-770.

Wiek, A.H.J., Melnick, R., van der Leeuw, S., Fukushi, K., Takeuchi, K., Farioli, F., Yamba, F., Blake, A., Geiger,C. and Kutter, R. 2015. Sustainability science in action: a review of the state of the field through case studies on disaster recovery, bioenergy,and precautionary purchasing. *Sustain Sci.*, 10 : 17–31.

Wikipedia. San Francisco: Wikipedia. 2012a Timeline of evolutionary history of life:http://en.wikipedia.org/wiki/Timeline of evolutionary history of life.

Wikipedia. San Francisco: Wikipedia, 2012b Pangaea. http://en.wikipedia.org/wiki/Pangaea.

Wikipedia. San Francisco: Wikipedia, 2012c. Gondwana. http://en.wikipedia.org/ wiki/Gondwana.

Wikipedia. San Francisco: Wikipedia, 2012d. Kumari Kandam.http://en.wikipedia.org/wiki/ Kumari_Kandam.

Wikipedia. San Francisco: Wikipedia, 2012e. Agastya. http://en.wikipedia.org/wiki/ Agastya.

Wikipedia, 2012f. Tamil Sangams.http://en.wikipedia.org/wiki/ Tamil_Sangams.

Wikipedia. San Francisco: Wikipedia, 2012g. Siddha medicine. http://en.wikipedia.org/wiki/ Siddha_medicine.

Wikipedia. San Francisco: Wikipedia. 2012h. Palm leaf manuscript. http://en.wikipedia.org/wiki/Palm- leaf_manuscript.

Wilson, E.O. (ed.). 1988. Biodiversity. Washington, DC, National Academy Press.

Wilson E.O. 1992. The diversity of Life. Questions of Life. Cambridge, MA, Harvard University Press.

Wisniewski, J., Frangne, N., Massonneau, A. and Dumas C. 2002. Between myth and reality: genetically modified maize, an example of a sizeable scientific controversy. *Biochimie* 84 : 1095–103.

Wood, D and Lenne, J.M. 2001. Nature's Fields: a neglected model for increasing food production. Outlook on Agriculture. pp. 161-170.

Zavaleta, E.S., Pasari, J.R., Hulvey, M.B. and Tilman, G.D. 2010. Sustaining multiple ecosystem functions in grassland communities requires higher biodiversity. *Proc. Natl. Acad. Sci. USA*, 107(4) : 1443-1446

Zisenis, M. 2015. Alien plant species : A real fear for urban ecosystems in Europe? *Urban Ecosyst.* 18(2) : 355-370.